Management Game Theory

Shaorong Sun · Na Sun

Management Game Theory

Shaorong Sun
Business School
University of Shanghai for Science
    and Technology
Yangpu district, Shanghai
China

Na Sun
Business School
University of Shanghai for Science
    and Technology
Yangpu district, Shanghai
China

ISBN 978-981-13-4559-3   ISBN 978-981-13-1062-1   (eBook)
https://doi.org/10.1007/978-981-13-1062-1

Jointly published with China Economic Publishing House, Beijing, China

The print edition is not for sale in China Mainland. Customers from China Mainland please order the print book from: China Economic Publishing House.

Printed on acid-free paper

This Springer imprint is published by the registered company Springer Nature Singapore Pte Ltd.
The registered company address is: 152 Beach Road, #21-01/04 Gateway East, Singapore 189721, Singapore

Preface

The content of management studies is very rich, with a large amount of theoretical and applied studies. As such, there is still much disagreement whether management studies are a subject in the liberal arts, science or engineering. I believe that whichever discipline management studies fall within, the most important mission of management studies is to be able to explain thoroughly the reason and mechanism that give rise to the various phenomena in practice, and to provide instruction and assistance for the various problems. In some respects, management studies are like medicine. In medical research, however abstract and abstruse, a 'model' is still useless if it cannot treat an illness or provide instruction and help for doctors. Similarly, the primary task in management research is to guide management practice by means of the profound theory.

As one of the main areas in management research, there is already a large amount of published work and teaching materials on game theory, providing abundant materials for studying game theory. However, these published work and teaching materials have one thing in common, in that they are mainly theoretical research. What readers find in these books are mostly abstract argument and lots of examples purely as numbers. Many people feel unable to apply them in actual management and research work.

The characteristic of this book is that while providing the necessary basics in game theory, it is mainly based on various game theory phenomena in management practice. Thus, it is convenient for the readers to see the profound game theory principle behind the various phenomena in management practice. Conversely, these game theory principles can also provide a degree of guidance for solving practical problems.

This book aims at being brief and concise, and is kept short, but the content is relatively comprehensive, including theories on non-cooperative games, cooperative games and evolutionary games and their related phenomena and questions in management practice.

The book is made up of two parts.

The first part focuses on the classical game theory. The content of this part comes mainly from several excellent reference books given at the end of this book, which was edited by the author of this book. The inclusion in this book of such content is mainly to enable the readers to have a relatively comprehensive understanding of the basics of game theory. In fact, these authors with their research and insights into their works greatly inspired the author of this book; theirs are the bases from which further research is carried out. Here, the author conveys sincere gratitude to the authors of those reference books.

The other part is the research of the author himself, some of which has already been published in academic journals. Different from the author's editing work in the first part, this part forms the author's 'writings' in this book. It is hoped that inclusion of these will inspire the readers to carry out further research.

This book is for teachers, students and cadres engaged in management studies and research, and it can also be used by postgraduate students as learning materials in game theory.

Research funding for this book: National Natural Science Foundation of China (71171134, 71771151); Shanghai World-class Discipline Project (S1201YLXK); Plateau Discipline Project of Shanghai (Management and Engineering).

Yangpu district, China   Shaorong Sun
                                                                                            Na Sun

Contents

1 The Fundamentals of Non-cooperative Games 1

1.1 Introduction to Game Theory 1

1.2 Three Essential Elements in Non-cooperative Games 4

1.3 Non-cooperation Game Model Form 5

    1.3.1 Representation of the Non-cooperative Game Matrix ... 6

    1.3.2 Representation of the Non-cooperative Game
Extensive Form 7

1.4 The Optimal Strategy (the Optimal Action) and the Worst
Strategy (the Worst Action) 8

    1.4.1 The Optimal Strategy (the Optimal Action) 8

    1.4.2 The Worst Strategy (the Worst Action) 9

    1.4.3 Relative Optimal Strategy 10

1.5 Nash Equilibrium and How to Find the Nash Equilibrium 11

    1.5.1 Nash Equilibrium 11

    1.5.2 Nash's Theorem and the Odd Number Theorem 12

1.6 How to Find the Nash Equilibrium in Static Games 13

    1.6.1 How to Find Pure Strategy Nash Equilibrium—Method
of Underlining Relative Optimal Strategy 13

    1.6.2 Finding Mixed Strategy Nash Equilibrium—Extremum
Method 14

    1.6.3 Finding Nash Equilibrium for Games with Continuous
Strategies—Extremum Method 17

1.7 Determining the Outcome of a Game with Multiple Nash
Equilibria 18

    1.7.1 Risk Control Principle 18

    1.7.2 The Principle of Pareto Optimum 19

1.8 Nash Equilibrium in Dynamic Games 20

2   Low Efficiency Caused by Non-cooperative Games   23
  2.1   Prisoner's Dilemma   23
  2.2   Competition and Cooperation for Duopoly—The Cournot
        Model and Cooperation Payoff   24
  2.3   The Irrational Equilibrium in the Tragedy of the Commons
        and Sun Shaorong's Fishing Model, and How the Number
        of People in a Collective Affects Efficiency   26
    2.3.1   An Introduction to the Tragedy of the Commons
            and Sun Shaorong's Fishing Model   26
    2.3.2   Sun Shaorong's Fishing Model   27
    2.3.3   Sun Shaorong's Non-rational Behaviour Equilibrium
            Point $B_0{}^*$ 30
    2.3.4   Collective Rationality Equilibrium Point B^{**}   31
    2.3.5   Comparisons of Non-rational Equilibrium Point $B_0{}^*$,
            Individual Rationality Nash Equilibrium Point B^*,
            and Collective Rationality Equilibrium Point B^{**}   32
    2.3.6   True Equilibrium Point—Depending on Society's
            Average Profit   34
    2.3.7   The Influence of the Quantity of Collective (n)
            on Equilibrium Point   34

**3   Free Ride, Adverse Selection, Moral Hazard and Separating
    Equilibrium**   37
  3.1   Free Ride Under Conditions of Symmetrical Information   37
  3.2   Asymmetric Information and Adverse Selection   39
    3.2.1   Asymmetric Information   39
    3.2.2   Adverse Selection   39
    3.2.3   Ways to Solve Adverse Selection   41
  3.3   Separating Equilibrium   42
  3.4   Asymmetric Information and Moral Hazard   49
  3.5   Some Approaches for Solving Moral Hazard   50

4   First-Move Advantage and Second-Move Advantage   53
  4.1   First-Move Advantage..............................   53
    4.1.1   First-Move Advantage in Discrete Behaviour.........   53
    4.1.2   First-Move Advantage in Continuous Behaviour   56
  4.2   Second-Move Advantage.............................   60
  4.3   Both First-Move Advantage and Second-Move Advantage
        May Be Present   61

5   Credible Commitment and Credible Threat in Games   63
  5.1   Credibility of Commitments and Threats   63
    5.1.1   Commitment and Threat.......................   63
    5.1.2   Credibility of Commitments and Threats   63

5.2   How to Increase the Credibility of Commitments
       and Threats   64
5.3   Both Parties in a Game Have no Credible Threat—The
       Chicken Game   65
5.4   The Market Entry Game—One Side Has no Credible
       Threat   65
5.5   Burning One's Bridges: One Side with Credible Threat   66

6   Coalitional Games   69
6.1   Coalitions   69
       6.1.1   Grand Coalition   69
       6.1.2   Number of Sub-coalitions   70
       6.1.3   The Number of All Coalitions   70
       6.1.4   Characteristic Vector   71
       6.1.5   Characteristic Function   71
       6.1.6   Definition of Cooperative Games   72
6.2   Imputation and the Core   73
       6.2.1   Collective Rationality—The Feasibility of the
               Allocation Scheme and the Player's Degree of
               Satisfaction   74
       6.2.2   Individual Rationality—Condition for an Individual
               to Join a Coalition   75
       6.2.3   The Imputation Set   75
       6.2.4   Small Coalition Rationality   76
       6.2.5   The Core   76
       6.2.6   The Problem that Exists When the Core Is the Solution
               to the Coalitional Game   77
6.3   The Shapley Value   81

7   Allocating Benefits in Coalitions   85
7.1   Allocating System Efficiency   85
7.2   The Problem of Unequal Status   86
       7.2.1   A Landlord and Tenant Cooperative Game   86
       7.2.2   The Problem of Asymmetric Pairing   88
7.3   The Question of Sets of Cost   89

8   Coalitions—Disintegration and Stability   93
8.1   Individual Rationality Causes a Cooperative Game
       to Become Confrontational and Diminished Payoffs   93
8.2   Small Coalition Rationality Causes a Cooperative Game
       to Become Confrontational and Diminished Payoffs   94
8.3   Individual Rationality Causes the Disintegration
       of a Coalition   96
8.4   Small Coalition Rationality Causes the Disintegration
       of a Coalition   97

    8.5   A Coalition's Stability   98
          8.5.1   Confrontations and a Coalition's Stability   98
          8.5.2   Definitions of Confrontation
                   and Counter-Confrontation   100
    8.6   Stability of Game Behaviour in Non-cooperative Games   100

9   Bottom Line for Negotiations and Solutions   103
    9.1   Overview of Negotiations and Negotiation Proportional
           Models   103
          9.1.1   An Overview of Negotiations   103
          9.1.2   Negotiation Proportional Models   103
    9.2   A General Model for Negotiations and an Objective
           Bottom Line   105
          9.2.1   A General Model for Negotiations   105
          9.2.2   The Objective Bottom Line in a Negotiation   106
    9.3   The Set of Solutions for a Successful Negotiation   108
    9.4   A Negotiation's Nash Product Solution   108

10 Evolution and Stability   115
    10.1   Evolutionarily Stable Point   115
    10.2   Mathematical Conditions for an Evolutionarily Stable Point . . .   119
    10.3   Follow-the-Crowd Game—the Evolutionary Equilibrium Point
           and the Stable Point of Evolutionary Equilibrium   119
    10.4   Evolutionarily Stable Point in a Central-Tendency
           Game—the Hawk-Dove Game   122
    10.5   Evolutionarily Stable Point and the Evolutionary Equilibrium
           Point in the Layabout Game   124

Bibliography   129

Synopsis

Based on the fundamentals of game theory, the game problems in this book are organised from various game phenomena in management practice. Thus, the readers can easily understand the profound game theory principles behind the management practice. Meanwhile, these game theory principles can also provide some guidance on solving practical problems.

This book is clear and concise, while the content is comprehensive, including theories on non-cooperative games, cooperative games and evolutionary games and the related phenomena and problems in management practice.

This book can be used as reading material for teachers, students and officials engaged in management studies and research, or as teaching material for post-graduate students studying game theory.

Chapter 1
The Fundamentals of Non-cooperative Games

1.1 Introduction to Game Theory

Game (*boyi* in Chinese) means playing chess in ancient China. Nowadays, it is mainly about choosing the most advantageous plan of action given the effect the opponent has on us. The theory on the study of games is called game theory.

In real life, there are broadly three main types of games.

The first type is when the object of the game is determined in advance, and the question concerning the game is to choose the most advantageous action in the given situation determined by the other player. This kind of game is called non-cooperative games.

Example 1.1 Price wars

In the products market, each enterprise works to promote the sale of their own products and to crowd out the sale of similar products by other enterprises. This is usually achieved through formulating lower prices. The problem is that everybody wants to increase the sale of their own products by lowering prices, which leads to a competition for price reduction, the so-called 'price war'. In this situation, the important question for each enterprise is how to formulate prices that are not higher than those of other enterprises and will not cause a serious loss to the enterprise. This is a classic question for non-cooperative games.

Non-cooperative games can be further divided into static games and dynamic games.

During the game, if each player chooses their own strategy without knowing the strategy chosen by other players, the game is called a static game. The well-known Prisoner's Dilemma is a classic static game.

If each player chooses their own strategy in a particular order, and if each player can observe the action and strategy taken by all the players before, the game is called a dynamic game.

© China Economic Publishing House and Springer Nature Singapore Pte Ltd. 2018
S. Sun and N. Sun, *Management Game Theory*,
https://doi.org/10.1007/978-981-13-1062-1_1

In a dynamic game, if the number of players is limited, and if the players continuously choose their strategy in a specified sequence, this kind of game is called repeated games.

Otherwise, if the players cannot repeat to choose a strategy during the whole game, it is called a non-repeated game. Non-cooperative games generally refer to non-repeated games unless stated otherwise.

The second type is cooperative games. The object for this type of games is also definite. The question is finding a scheme for distributing payoffs obtained in cooperation which outweigh that of cost, so that every player feels that the benefits for cooperation are greater than non-cooperation. This kind of games is called cooperative games.

Example 1.2 Payoff distribution

Let us assume there are four players 1, 2, 3, 4 discussing whether or not to cooperate in starting an enterprise: If all 4 cooperate, the enterprise can earn up to 1,000,000 yuan annually; if 1 and 2 cooperate, then the enterprise's annual income will be 100,000 yuan; if 1 and 4 cooperate, the annual income can reach 500,000 yuan; if 2 and 3 cooperate, the annual income will be 300,000 yuan; if 3 and 1 cooperate the annual income will be 600,000 yuan; if 4 and 2 cooperate, the annual income will be 0; if 4 and 3 cooperate, the annual income will be 30,000 yuan; if each work by themselves, then the annual payoff for each of them will be 0.

Now, the 4 of them want to find a reasonable scheme for payoff distribution so that each and everyone of the 4 will be willing to cooperate in order to produce a good outcome of 1,000,000 yuan payoff. Clearly, if the payoff as any sub-group of fewer than 4 is greater than that of the payoff as a 'large' group, then those in the sub-group will not be willing to participate in the 4-person 'large' group. This is a classic cooperative game problem.

The third type is evolutionary games. The biggest difference with the two types before is that the object of this type of games is not definite. Usually, in a very large group of players there is a certain probability of likelihood that each player will meet another player of a certain 'type' (i.e. a player with a certain game strategy). In this situation, each player choosing different strategies can result in different payoff for themselves. In this situation, since the object of the game is indefinite, each player cannot know which strategy will be more advantageous. However, if the game repeats continuously, then each player will gradually 'learn' what strategy is more likely to bring better payoff. In this way, for the large group everyone will learn to continuously adjust their own strategy, until the ratio of different types of players stabilises, that is, this ratio does not change anymore. This is the process of the evolutionary game. The problem for study by the evolutionary game theory is to predict the direction of change to the ratio in the various types of players and the end point of that change (when the ratio stabilises).

Games as a social phenomenon have always existed in society. The idea of game theory emerged early on in society. *The Art of War*, a book in China over two thousand years ago, exhibited many ideas on game theory.

In 1712 James Waldegrave suggested that the minimax mixed strategy can be used to play the best game. This says that if each of our own strategy leads to many different results depending on the other player's choice of strategy, then we should first find the least advantageous result from our own strategy (i.e. the lowest value of payoff from our own strategies). Then compare these values of low payoff, and select the strategy that gives the highest result out of all these low values. This method can effectively prevent the risk of loss encountered with the strategy.

In 1838 the French economist A. A. Cournot used game ideas to analyse the problem of equilibrium quantity for competitive oligopoly in economic management. Using the reaction function, a profit function is derived which finds the game equilibrium, i.e. the equilibrium quantity.

During the 1920s the French mathematician Borel proposed the mixed strategy concept that reflects the phenomenon of uncertain moves in the process of games.

In 1944, von Neumann and Morgenstern jointly published the *Theory of Games and Economic Behaviour*. This publication is regarded as the beginning of a more systematic theory on games.

By the 1950s Gillies and Shapley proposed the concept of the 'Core' in cooperative games. This is an important development in cooperative game theory.

In the area of non-cooperative games, Nash came forward with the concept of the Nash Equilibrium and the Equilibrium Existence Theorem which states that for any finite number of players with finite strategies in non-cooperative games, if a mixed strategy is considered, then there is at least one equilibrium point. At the same time, Tucker came up with a well-known non-cooperative game example, the Prisoner's Dilemma, in 1950.

In 1965 Reinhard Selten focused on the question of the Nash Equilibrium in dynamic games and came forward with the subgame perfect Nash equilibrium concept.

In the 1960s game theory produced a very important branch: Evolutionary game. This branch is able to explain biological evolution. For example, in 1960 Lewontin began to use game theory to explain biological evolution.

In 1973 Maynard Smith and Price proposed the important fundamental concept in the evolutionary game theory: Evolutionary stability.

In 1982 John Maynard Smith published *Evolution and the Theory of Games*, a text regarded as a classic in the field of evolutionary game. Afterwards, Taylor and Jonker put forward the evolutionary dynamics equation which further established this theory. It also produced a lot of applied research. For example, people employ evolutionary games to analyse how social systems change over time, stock market trends, how consumers select brands, the formation of social customs, etc.

1.2   Three Essential Elements in Non-cooperative Games

There are three essential elements in non-cooperative games.

One essential element is the player. Some Chinese publications on games call this the player, the game player, etc. Participants are the main body of a game, independent in making judgement and decision. In a game there are at least two players, otherwise it cannot constitute a game. Of course, there can be three, four players, etc. In the evolutionary game, there is usually an infinite number of players.

Where there are 2 players, it is called a 2-person game, 3 players, a 3-person game, etc.

In publications of game theory, all the participants constitute a set of player. As N indicated n players form a set, the ith player is usually called 'player i'. For example, in the situation of $n=3$, the three players can be represented as player 1, player 2, player 3. Where it will not cause confusion, it can directly be said of 1, 2, 3.

The second essential element is the strategy or the action chosen by the player. Of these, strategy is with regularity. For instance: 'If the opponent attacks I will counter-attack' is a strategy; 'if the opponent attack I will yield' is also a strategy. However, action is purely a counter measure. For example, when the opponent comes to grab the enterprise's market share, one action is 'to counter-attack by lowering the price of our products'; but 'acceptance, no action taken' is also an action.

In many publications of game theory, where strategy and action are unified, the term 'strategy' is used because 'action' can be interpreted as a kind of simple strategy, or as a strategy without preconditions.

During a game, each player usually has many strategies to choose from. Therefore the strategies available for each player is called a strategy set. Generally s_i indicates a chosen strategy of player i, whereas S_i is the strategy set of player i.

The Cartesian product of the strategy sets of all the players $S = X_{i\in N} S_i$ is called the strategy space. Thus, regarding all the players, if everyone has chosen a specific strategy, then the strategy vector $s = (s_1, s_2, \ldots, s_i, \ldots, s_n)$ is a point in the strategy space $S = X_{i\in N} S_i$. Each point in the strategy space represents a play of the game. This is formed after every player has chosen their own strategy, and as such is a description of the chosen strategy of all the players.

In game theory there are two types of strategies. One is a pure strategy, and the other is a mixed strategy.

A pure strategy refers to the strategy a player must choose out of his strategy set when faced with all the strategies to choose from (i.e. his strategy set). In other words, of the various strategies in the strategy set, it is either chosen by the player or it is not chosen. Represented as that player's strategy vector, it is:

Assume the strategy set of player i is $S_i = \{s_{i1}, s_{i2}, \ldots, s_{im_i}\}$, of this, m_i is the number of elements of the strategy set $S_i = \{s_{i1}, s_{i2}, \ldots, s_{im_i}\}$.

Again, suppose player i in choosing a strategy from his own strategy set, the vector resulting from the probability in using each of the strategy in the strategy set $S_i = \{s_{i1}, s_{i2}, \ldots, s_{im_i}\}$ (abbreviated as player i's strategy probability vector) is $P_i = \{p_{i1}, p_{i2}, \ldots, p_{im_i}\}$, if $P_i = \{p_{i1}, p_{i2}, \ldots, p_{im_i}\}$ meets the conditions

$$p_{ij} = \begin{cases} 1, & \text{When } s_{ij} \text{ is adopted} \\ 0, & \text{When } s_{ij} \text{ is not adopted} \end{cases}, \text{ as well as } \sum_{j=1}^{m_i} p_{ij} = 1, \text{ then the strategy}$$

adopted by player i is a pure strategy.

In a game, if all the players adopt pure strategies, then the game is called a pure strategy game.

Regarding the strategy probability vector of player i $P_i = \{p_{i1}, p_{i2}, \ldots, p_{im_i}\}$, if $P_i = \{p_{i1}, p_{i2}, \ldots, p_{im_i}\}$, $0 \le p_{ij} \le 1$, and also $\sum_{j=1}^{m_i} p_{ij} = 1$, then the strategy adopted by player i is a mixed strategy. In other words, in adopting a mixed strategy, the probability of the player choosing any element in the strategy set is between 0 and 1 (including 0 and 1). In such situation, there is a certain amount of uncertainty as to the strategy to be adopted by the player.

In a game, if all the players adopt mixed strategies, then the game is called a mixed strategy game.

The children's game of 'rock, paper, scissors' is a kind of mixed strategy game. There is a lot of uncertainty beforehand as to whether the other player will make the sign for 'rock', 'paper' or 'scissors'.

Broadly, pure strategy is a special case of mixed strategy; it occurs when the probability p_{ij} of the mixed strategy is the endpoint value (0 or 1).

The third essential element is payoff; some Chinese game publications also call it benefit or payout. It refers to the reward received after the player has chosen a certain strategy or action. It should be pointed out that some payoffs as expressed in functions often use the 'utility' concept so as to more accurately reflect the true reward for the player. In fact, since many payoffs are not economic indicators but some kind of social payoffs, such as winning a war, enjoying some prestige, promotion at work, etc., and since the players are under different circumstances and will feel differently to the same reward, therefore the concept of 'utility' can more accurately express the payoff of the game.

Generally speaking, the payoff for player i within the game's payoff is expressed as u_i. Thus, the payoff for all players also from a payoff vector $U = (u_1, u_2, \ldots, u_i, \ldots, u_n)$. It is actually an n-dimensional space called a payoff space.

With the three essential elements defined, a game can be expressed as $G = (N, S, U)$. Among this, N is the set of players, S is the strategy space, and U is the payoff space.

1.3 Non-cooperation Game Model Form

There are two forms when describing non-cooperative games—one is the matrix form, the other is the extensive form.

1.3.1   Representation of the Non-cooperative Game Matrix

The matrix is also called the payoff matrix. It shows, in table form, as a way to describe non-cooperative games, the payoffs of the various players under different combinations of strategies. Some publications of game theory sometimes call the game matrix as the 'normal-form game'.

The matrix is only suitable for describing the static game.

For a game with only two players, a 2-dimensional table form can be used, with one player's strategy runs horizontally (called the row strategy), because each row in the table represents a strategy of that player. The other player's strategy runs vertically (called column strategy), because each column on the table represents a strategy of that player. The corresponding payoffs from the various strategy profiles for both players are written in the table. For instance, for the well-known Prisoner's Dilemma, the matrix is (Table 1.1):

Table 1.1 Prisoner's dilemma matrix

	B denies	B confesses
A denies	Serve a 1-year sentence, serve a 1-year sentence	Serve a 10-year sentence, immediately released
A confesses	Immediately released, serve a 10-year sentence	Serve an 8-year sentence, serve an 8-year sentence

This game is about the police arresting the suspects A and B. Because the police interrogate the two separately, a non-cooperative game began between the two suspects. The available behaviour sets for A and B are 'deny, confess'. The payoffs are in each box—A's payoff (payout) is on the left, B's payoff (payout) is on the right. In Table 1.1, A's strategy is the row, and B's strategy is the column.

Table 1.2 is for 3×2 strategy game matrix.

Table 1.2 3×2 strategy game matrix

	Player 2's strategy 1	Player 2's strategy 2
Player 1's strategy 1	0, 3	2, 6
Player 1's strategy 2	2, 1	4, 2
Player 1's strategy 3	6, 2	3, 1

Regarding the number of tables in the matrix form, when there are two players, there is only one table. If there are more than two players, then the number of tables in the matrix will increase. If the number of players is $n \geq 3$, the number of tables in the matrix is t, then $t = S_3 \times S_4 \times \cdots S_n$; among this, S_i is the number of strategies in the strategy set of player i. So, for $n \geq 3$, the number of tables is the product, after the third player, of the strategy number in each of the player's strategy set. It can also

Table 1.3   a One of $4 \times 3 \times 2$ strategy game matrices (when player 3 adopts strategy 1)

	Player 2's strategy 1	Player 2's strategy 2	Player 2's strategy 3
Player 1's strategy 1	0, 3, 2	2, 6, 3	1, 0, 8
Player 1's strategy 2	2, 1, 3	4, 2, 4	3, 1, 2
Player 1's strategy 3	6, 2, 2	3, 1, 1	4, 0, 3
Player 1's strategy 4	3, 1, 1	2, 2, 3	2, 3, 2

Table 1.3   b One of $4 \times 3 \times 2$ strategy game matrices (when player 3 adopts strategy 2)

	Player 2's strategy 1	Player 2's strategy 2	Player 2's strategy 3
player 1's strategy 1	2, 1, 0	1, 3, 2	2, 1, 3
player 1's strategy 2	1, 2, 1	2, 2, 1	1, 2, 0
player 1's strategy 3	3, 4, 2	0, 3, 2	3, 1, 2
player 1's strategy 4	0, 5, 4	3, 1, 4	4, 2, 3

be treated as the number of Cartesian product 'points' from the strategy set of each player after the third player.

For example, when player 1's strategy is 4, player 2's strategy is 3, player 3's strategy number is 2, then the game matrix is Table 1.3a and 1.3b.

As it shows in Table 1.3b, when player 1 adopts strategy 4, player 2 adopts strategy 3, player 3 adopts strategy 2, the payoff for the three players are respectively 4, 2 and 3 (see shaded area in Table 1.3b).

1.3.2   Representation of the Non-cooperative Game Extensive Form

The extensive form game is called Extensive Game Mold in English. It is the main form to describe the dynamic game. The representation of the extensive form game is often called a game tree in some game theory publications.

In the extensive form nodes and edges form a branching out tree diagram. A game tree usually has an initial node called the root of the game tree. Then it grows continuously downwards or to the right. As each node grows, there is at least one edge (usually there are at least two or more edges). Each node represents the point in time where a certain player can choose a specific strategy from his own strategy set. The number of edges from this node extending downwards or to the right is the number of the strategies from that player's strategy set at this point in time.

It should be pointed out that the nodes at the lowest or furthest point to the right (whether it is furthest down or right is dependent on the direction of the game tree as it grows) where edges stop extending are called terminal nodes. At a terminal node, the player's payoff is usually written down to indicate the end of the game.

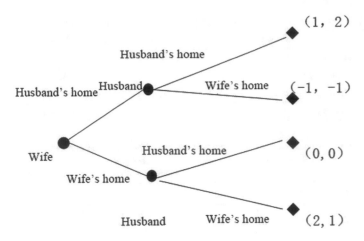

Fig. 1.1 The wife makes the first move in a husband and wife game

Figure 1.1 describes a young couple working away from home. The problem of whether to return to the husband's home or the wife's home for the Chinese New Year forms a dynamic game tree. The husband wants to spend the festival with his family, while the wife wants to spend it with her family, but the husband and wife do not want to celebrate the Chinese New Year separately (which is the worst outcome for both). If both of them go to the wife's home for the Chinese New Year, then the wife's payoff is 2, while the husband's payoff is 1; if they both go to the husband's home for the Chinese New Year, then the wife's payoff is 1, while the husband's payoff is 2; if they separate and go back to their own homes respectively, then the payoff for both is 0. Figure 1.1 shows a dynamic game where the wife makes the first move.

1.4   The Optimal Strategy (the Optimal Action) and the Worst Strategy (the Worst Action)

1.4.1   The Optimal Strategy (the Optimal Action)

In non-cooperative games, sometimes this situation emerges for a player: Regardless of what strategy (action) is chosen by other players, when one chooses a certain strategy (action) the payoff is always bigger than that from one's other strategies put together. This strategy is that player's optimal strategy.

Let's assume that with regards to game $G = (N, S, U)$, S_{-i} shows the $n - 1$ dimensional strategy space formed by subtracting player i's strategy set S_i from the strategy set $S = \times_{i \in N} S_i$. $s_{-i} \in S_{-i}$ represents any node in S_{-i} (i.e. any strategy profile that does not include player i's strategy); s_i^* represents one of player i's specific strategy; $\{S_i - s_i^*\}$ represents the strategy set formed by subtracting s_i^* from player i's strategy set. So for any $s_i \in \{S_i - s_i^*\}$ there is:

$$u(s_i^*, s_{-i}) \geq u(s_i, s_{-i})$$

Then s_i^* is player i's optimal strategy.

For instance, in the Prisoner's Dilemma, for A, the result of choosing to 'confess' is always better than choosing to 'deny' whether B chooses to 'deny' or 'confess'. Therefore to 'confess' is optimal strategy for A. Likewise, 'confess' is also B's optimal strategy.

1.4.2 The Worst Strategy (the Worst Action)

The worst strategy is opposite to optimal strategy. Sometimes this happens to one of the players: Regardless of what strategy (action) is chosen by other players, when one chooses a certain strategy (action) the payoff is always smaller than that from one's other strategies put together. This strategy is that player's worst strategy.

Let's assume that with regards to game $G = (N, S, U)$, S_{-i} shows the $n - 1$ dimensional strategy space formed by subtracting player i's strategy set S_i from strategy space $S = \times_{i \in N} S_i$. $s_{-i} \in S_{-i}$ shows any node in S_{-i} (i.e. any strategy profile that does not include player i's strategy). s_i^- shows one of player i's specific strategy. $\{S_i - s_i^*\}$ shows player i's strategy set S_i formed without s_i^-. So for any $s_i \in \{S_i - s_i^*\}$ there is:

$$u(s_i^-, s_{-i}) \leq u(s_i, s_{-i})$$

Then s_i^* is player i's worst strategy.

For instance, in the Prisoner's Dilemma, for A, whether B chooses to 'deny' or 'confess', if A chooses to 'deny', his outcome is always worse than choosing to 'confess'. Therefore to 'deny' is the worst strategy for A. Likewise, 'deny' is also B's worst strategy (Table 1.1).

Table 1.1 Prisoner's dilemma matrix

	B denies	B confesses
A denies	Serve a 1-year sentence, serve a 1-year sentence	Serve a 10-year sentence, immediately released
A confesses	Immediately released, serve a 10-year sentence	Serve an 8-year sentence, serve an 8-year sentence

1.4.3 Relative Optimal Strategy

Compared to relative optimal strategy, a lesser case is with regards to a certain strategy profile of other players (note that this is others' strategy profile, i.e. not including the player referred to), when the payoff of that player's chosen strategy (action) is always greater or equal to that of his any other strategy. This strategy is that player's optimal strategy with regards to the strategy profile of other players.

For instance, let us suppose that the game matrix of players A and B's non-cooperative game is as represented as in Table 1.4.

Table 1.5 indicates player A's relative optimal strategy. It can be seen from this table that for player A, if B chooses strategy 1, then A's relative optimal strategy is strategy 2; if B chooses strategy 2, then A's relative optimal strategy is strategy 1, of which the payoff from A's relative optimal strategy is underlined in Table 1.5.

Table 1.6 indicates player B's relative optimal strategy. For player B, if A chooses strategy 1, then B's relative optimal strategy is strategy 1; if A chooses strategy 2, then B's relative optimal strategy is strategy 2, of which the payoff from B's relative optimal strategy is underlined in Table 1.6.

Table 1.4 An example of a two-person non-cooperative game

	B's strategy 1	B's strategy 2
A's strategy 1	2, 3	5, 1
A's strategy 2	3, 1	2, 4

Table 1.5 Player A's relative optimal strategy

	B's strategy 1	B's strategy 2
A's strategy 1	2, 3	$\underline{5}$, 1
A's strategy 2	$\underline{3}$, 1	2, 4

Table 1.6 Player B's relative optimal strategy

	B's strategy 1	B's strategy 2
A's strategy 1	2, $\underline{3}$	5, 1
A's strategy 2	3, 1	2, $\underline{4}$

1.5 Nash Equilibrium and How to Find the Nash Equilibrium

1.5.1 Nash Equilibrium

Nash Equilibrium is a very important term in non-cooperative games, named after its proponent John Nash.

Nash Equilibrium is a stable combination of chosen strategy by all the players in a non-cooperative game. Here, the strategy of each player is the relative optimal strategy relative to other players' strategy profile. In other words, when Nash Equilibrium is reached, for any players, when other players' current strategy remains unchanged, if that player changes his own current strategy, it will lead to a reduced payoff for himself (when his strategy is strictly optimal), or at least it will not lead to increased payoff (when his strategy is weakly optimal).

Nash Equilibrium came from John Forbes Nash Jr.'s doctoral dissertation entitled 'Non-cooperative Game' (1950). This dissertation was published as two papers: 'Equilibrium Points in N-person Games' (1950), and 'Non-cooperative Games' (1951). These papers presented the generic solution that exists in non-cooperative games of any number of players (what scholars discussed before Nash was basically two-person zero-sum games). This solution was later called Nash Equilibrium.

The formal definition for Nash Equilibrium:

Let us suppose that $s^* = (s_1^*, \ldots, s_n^*)$ is a node in the strategy space S of the game $G = (N, S, U)$ (i.e. the strategy profile of n players), if, for each player i there is:

$$u_i(s_1^*, \ldots, s_{i-1}^*, s_i^*, s_{i+1}^*, \ldots, s_n^*) \geq u_i(s_1^*, \ldots, s_{i-1}^*, s_i, s_{i+1}^*, \ldots, s_n^*),$$

then $s^* = (s_1^*, \ldots, s_n^*)$ is called a Nash Equilibrium of game $G = (N, S, U)$.

Note that in $s^* = (s_1^*, \ldots, s_n^*)$ the strategy of each of the player can be pure strategy or mixed games.

Example 1.3 Prisoner's Dilemma and Nash Equilibrium

For the convenience of comparison, all the payoffs in Table 1.1 of the Prisoner's Dilemma are represented in numbers (number of years in prison) (Table 1.7). Note that the bigger the number, the longer the prison sentence, and therefore the two players are seeking the smallest possible number.

There is only one pure strategy Nash Equilibrium in this game, i.e. (A confesses, B confesses). It can be seen from Table 1.7 that, whether B chooses to 'deny' or 'con-

Table 1.7 Prisoner's dilemma matrix in numbers

	B denies	B confesses
A denies	1, 1	10, 0
A confesses	0, 10	8, 8

Table 1.8 Another example of Nash equilibrium

	B strategy 1	B strategy 2
A strategy 1	4, 2	5, 6
A strategy 2	6, 8	3, 2

Table 1.4 An example of a two-person non-cooperative game

	B's strategy 1	B's strategy 2
A's strategy 1	2, 3	5, 1
A's strategy 2	3, 1	2, 4

fess', A's optimal strategy is to always to 'confess', and as for B, likewise, whether A chooses to 'deny' or to 'confess', B's optimal strategy is always to 'confess'.

Example 1.4 Another example of Nash Equilibrium

Table 1.8 is another example of a two-person non-cooperative game. In this example we assume that the number for the payoff is the bigger the better. From Table 1.8 it is found that there are two pure strategy Nash Equilibria: (A strategy 1, B strategy 2) and (A strategy 2, B strategy 1).

Example 1.5 No pure strategy Nash Equilibrium

It can be seen from Table 1.4 that if A chooses strategy 1, then B should choose strategy 1; whereas if B has chosen strategy 1, then A will choose strategy 2. After A has chosen strategy 2, B will choose strategy 2; after B has chosen strategy 2, A will return to strategy 1. Thus begins a new cycle. Therefore the game shown in Table 1.4 does not contain a Nash Equilibrium of pure strategy.

From the several examples above, it can be seen that for a non-cooperative game of limited players with limited strategy, it sometimes does not contain a pure strategy Nash Equilibrium, sometimes contains only one Nash Equilibrium, sometimes more than one Nash Equilibrium.

1.5.2 Nash's Theorem and the Odd Number Theorem

The number of Nash Equilibrium differs in the three examples above. Therefore, the natural question is, under normal circumstances for a non-cooperative game $G = (N, S, U)$, how many Nash Equilibria are there?

Regarding this, Nash gave us the answer in 1950, namely the Nash's Theorem.

Nash's Theorem: In a situation where the number of players is limited and the pure strategy set of each player is limited, if the pure strategy set of each player is expanded as a mixed strategy set, then in the non-cooperative game $G = (N, S, U)$ there is at least one Nash Equilibrium.

Nash's Theorem states that with regards to non-cooperative games with limited players, if mixed strategy is considered, then at least one Nash Equilibrium can be found.

Further, Professor Robert Wilson proved in 1971 that: The number of Nash Equilibrium is limited in almost all static games of limited strategy, and this number is an odd number. This is the well-known odd number theorem.

1.6 How to Find the Nash Equilibrium in Static Games

1.6.1 How to Find Pure Strategy Nash Equilibrium—Method of Underlining Relative Optimal Strategy

How do we find the Nash Equilibrium point if given a payoff matrix in a static game? In *Introduction to Game Theory* written by Wan Zeke and Li Jie, a very simple way of finding the Nash Equilibrium was given—underlining the relative optimal strategy.

Aiming at each player, this method finds one's relative optimal strategy given the various strategy profiles of other players, and draws lines under the figure of one's own payoff as a result of one's relative optimal strategy. In the payoff matrix, if in a box the payoff figures of all the players have lines, then the players' strategies in that box form a Nash Equilibrium.

Clearly, underlining the relative optimal strategy is only suitable for finding pure strategy Nash Equilibrium.

For instance for the non-cooperative game in Table 1.8, the steps to underline the relative optimal strategy are:

Table 1.8 An instance of the non-cooperative game

	B strategy 1	B strategy 2
A strategy 1	4, 2	5, 6
A strategy 2	6, 8	3, 2

First, find A's relative optimal strategy:

Let us suppose that B adopts strategy 1, then A's relative optimal strategy is strategy 2; if B adopts strategy 2, then A's relative optimal strategy is strategy 1. Thus in Table 1.8, the result of the lines are as in Table 1.9:

Table 1.9 Finding A's relative optimal strategy

	B strategy 1	B strategy 2
A strategy 1	4, 2	<u>5</u>, 6
A strategy 2	<u>6</u>, 8	3, 2

Table 1.10 Finding B's relative optimal strategy

	B strategy 1	B strategy 2
A strategy 1	4, 2	<u>5</u>, <u>6</u>
A strategy 2	<u>6</u>, <u>8</u>	3, 2

Then, find B's relative optimal strategy. Let us suppose that A adopts strategy 1, then B's relative optimal strategy is strategy 2; if A adopts strategy 2, then B's optimal strategy is strategy 1. Thus in Table 1.9, the result of the lines are as in Table 1.10.

Looking at Table 1.10 two Nash Equilibria can be observed: (strategy 1, strategy 2), (strategy 2, strategy 1).

1.6.2 Finding Mixed Strategy Nash Equilibrium—Extremum Method

Drawing lines is only suitable for finding the pure strategy Nash Equilibrium. Then what about finding the mixed strategy Nash Equilibrium?

Table 1.11 shows the non-cooperative game matrix of 'rock, paper, scissors' played by two children. In this, the payoff is 1 for the winner, −1 for the loser, and 0 for both if it is a tie.

It can be seen from Table 1.11 that this is a zero-sum game, i.e. the sum of the winnings by A and B is zero. This is a very competitive non-cooperative game, namely that if one side obtains a payoff, the other side must lose.

Let's first try to find the pure strategy Nash Equilibrium by underlining the optimal strategy. The lines are as shown in Table 1.12:

Looking at Table 1.12, it can be seen that none of the boxes contain any lines under the payoff for the two sides in this game. Therefore there is no pure strategy Nash Equilibrium in this game.

Table 1.11 Rock-paper-scissors game

	B rock	B scissors	B paper
A rock	0, 0	1,−1	−1, 1
A scissors	−1,1	0, 0	1,−1
A paper	1,−1	−1, 1	0,0

Table 1.12 Underlining the optimal strategy

	B rock	B scissors	B paper
A rock	0, 0	<u>1</u>,−1	−1,<u>1</u>
A scissors	−1,<u>1</u>	0, 0	<u>1</u>,−1
A paper	<u>1</u>,−1	−1,<u>1</u>	0,0

Let us now see if this game contains a mixed strategy Nash Equilibrium. For this, let us suppose:

The probability of A plays 'rock', 'scissors', 'paper' in turn is: $p_{11}, p_{12}, 1 - (p_{11} + p_{12})$

The probability of B plays 'rock', 'scissors', 'paper' in turn is: $p_{21}, p_{22}, 1 - (p_{21} + p_{22})$

When A adopts the strategy for 'rock', his expected payoff is:

$$u_{11} = 0 \times p_{21} + 1 \times p_{22} + (-1) \times p_{23} = p_{22} - p_{23}$$

When A adopts the strategy for 'scissors', his expected payoff is:

$$u_{12} = (-1) \times p_{21} + 0 \times p_{22} + 1 \times p_{23} = -p_{21} + p_{23}$$

When A adopts the strategy for 'paper', his expected payoff is:

$$u_{13} = 1 \times p_{21} + (-1) \times p_{22} + 0 \times p_{23} = p_{21} - p_{22}$$

When B adopts the strategy for 'rock', his expected payoff is:

$$u_{21} = 0 \times p_{11} + 1 \times p_{12} + (-1) \times p_{13} = p_{12} - p_{13}$$

When B adopts the strategy for 'scissors', his expected payoff is:

$$u_{22} = (-1) \times p_{11} + 0 \times p_{12} + 1 \times p_{13} = -p_{11} + p_{13}$$

When B adopts the strategy for 'paper', his expected payoff is:

$$u_{23} = 1 \times p_{11} + (-1) \times p_{12} + 0 \times p_{13} = p_{11} - p_{12}$$

Since it is supposed that both children use mixed strategy (i.e. sometimes 'rock', sometimes 'scissors', sometimes 'paper', the choices for each child are the probability of playing these gestures), therefore:

A's expected payoff is:

$$
\begin{aligned}
u_1 &= p_{11}u_{11} + p_{12}u_{12} + p_{13}u_{13} \\
&= p_{11}(p_{22} - p_{23}) + p_{12}(-p_{21} + p_{23}) + p_{13}(p_{21} - p_{22}) \\
&= p_{11}(p_{22} - p_{23}) + p_{12}(-p_{21} + p_{23}) + (1 - p_{11} - p_{12})(p_{21} - p_{22}) \\
&= p_{11}(2p_{22} - p_{23} - p_{21}) + p_{12}(-2p_{21} + p_{23} + p_{22}) + (p_{21} - p_{22})
\end{aligned}
$$

B's expected payoff is:

$$
\begin{aligned}
u_2 &= p_{21}u_{21} + p_{22}u_{22} + p_{23}u_{23} \\
&= p_{21}(p_{12} - p_{13}) + p_{22}(-p_{11} + p_{13}) + (1 - p_{21} - p_{22})(p_{11} - p_{12})
\end{aligned}
$$

$$= p_{21}(2p_{12} - p_{13} - p_{11}) + p_{22}(-2p_{11} + p_{13} + p_{12}) + (p_{11} - p_{12})$$

Then, seek the optimal strategies for both A and B.

Let us first seek the partial derivatives of A's and B's payoff functions respectively:

$$\frac{\partial u_1}{\partial p_{11}} = 2p_{22} - p_{23} - p_{21} = 2p_{22} - 1 + p_{22} + p_{21} - p_{21} = 3p_{22} - 1,$$

$$\frac{\partial u_1}{\partial p_{12}} = -2p_{21} + p_{23} + p_{22} = -2p_{21} + 1 - p_{21} - p_{22} + p_{22} = 1 - 3p_{21}$$

$$\frac{\partial u_2}{\partial p_{21}} = 2p_{12} - p_{13} - p_{11} = 2p_{12} - 1 + p_{12} + p_{11} - p_{11} = 3p_{12} - 1$$

$$\frac{\partial u_2}{\partial p_{22}} = -2p_{11} + p_{13} + p_{12} = -2p_{11} + 1 - p_{11} - p_{12} + p_{12} = 1 - 3p_{11}$$

Solving the simultaneous equations:

$$\begin{cases} \dfrac{\partial u_1}{\partial p_{11}} = 3p_{22} - 1 = 0 \\[2mm] \dfrac{\partial u_1}{\partial p_{12}} = 1 - 3p_{21} = 0 \\[2mm] \dfrac{\partial u_2}{\partial p_{21}} = 3p_{12} - 1 = 0 \\[2mm] \dfrac{\partial u_2}{\partial p_{22}} = 1 - 3p_{11} = 0 \end{cases}$$

We have:

$$p_{11} = \frac{1}{3}, \quad p_{12} = \frac{1}{3}, \quad p_{13} = 1 - p_{11} - p_{12} = \frac{1}{3};$$

$$p_{21} = \frac{1}{3}, \quad p_{22} = \frac{1}{3}, \quad p_{23} = 1 - p_{21} - p_{22} = \frac{1}{3}$$

i.e. the Nash Equilibrium of the two children is $[(\frac{1}{3}, \frac{1}{3}, \frac{1}{3}), (\frac{1}{3}, \frac{1}{3}, \frac{1}{3})]$, that is the probability of A playing 'rock', 'scissors', 'paper' is one-third, and the probability of B playing 'rock', 'scissors', 'paper' is also one-third.

Think carefully about this, and the Nash Equilibrium makes sense: So long as the other player plays a certain gesture (such as the 'rock') with a greater probability, a corresponding gesture can be found immediately (such as playing 'paper') in order that one can win more frequently. Therefore optimal strategy is the same probability for every gesture. Thus, the other player has no opportunity to gain the upper hand.

1.6.3 *Finding Nash Equilibrium for Games with Continuous Strategies—Extremum Method*

In static non-cooperative games, if a player's strategy set is continuous, for example strategy set S_i is continuous, that is S_i is made up of infinitely continuous strategies. How do we find the Nash Equilibrium in such situations?

The extreme value method can be used to find the Nash Equilibrium.

Below, the Cournot model is used as an example to explain this method.

The Cournot model describes the process of how two manufacturers who produce a homogeneous product reach equilibrium in the game. The conditions are:

A certain product is manufactured by two enterprises 1 and 2; the production cost of that product is zero.

The market demand price function is $P = a - (q_1 + q_2)$. Within this, P is the product's selling price, a is the constant, q_1 and q_2 are the outputs of enterprises 1 and 2. The two enterprises both accurately understand the demand curve in that market.

Under non-cooperative game conditions, the two enterprises will first estimate the other's output, and then determine their own output that will give the enterprise the biggest profit.

More importantly, when the two enterprises decide on their own output, it must be done simultaneously without the situation of one enterprise determining its output before the other.

Let us suppose that:

q_1 is enterprise 1's output, q_2 is enterprise 2's output, u_1 is enterprise 1's profit, u_2 is enterprise 2's profit, then:

$$u_1 = q_1 P = q_1[a - (q_1 + q_2)]$$
$$u_2 = q_2 P = q_2[a - (q_1 + q_2)]$$

Seek the optimal output for enterprise 1 and 2:

$$\frac{\partial u_1}{\partial q_1} = a - (q_1 + q_2) - q_1 = a - q_2 - 2q_1 = 0$$

$$\frac{\partial u_2}{\partial q_2} = a - (q_1 + q_2) - q_2 = a - q_1 - 2q_2 = 0$$

And the optimal output for enterprise 1 and 2 are:

$$q_1^* = \frac{1}{2}(a - q_2)$$
$$q_2^* = \frac{1}{2}(a - q_1)$$

Everything being equal between the two enterprises, solving the above simultaneous equations will arrive at the Nash Equilibrium of the simultaneous decisions on output by the two enterprises:

$$q_1^* = q_2^* = \frac{a}{3}$$

Therefore, the Nash Equilibrium in a Cournot game is: $(\frac{a}{3}, \frac{a}{3})$

1.7 Determining the Outcome of a Game with Multiple Nash Equilibria

In many situations, there are often many Nash Equilibria in a non-cooperative game. In such situation, which Nash Equilibrium is the most likely outcome of the game?

In a situation of many Nash Equilibria, the most likely game outcome is actually decided by the views of the game players, i.e. their principle in choosing the final outcome.

1.7.1 Risk Control Principle

Risk control principle means that in choosing one's strategy, the player will first choose the strategy with the least risk to oneself. See the game shown below in Table 1.13.

Table 1.13 shows the game as: the products manufactured by enterprises A and B are complementary. The strategy for both sides can be divided into three situations:

If enterprises A and B act in concert with each other, i.e. the strategy profiles of A and B are (A acts in concert, B acts in concert), then the payoffs for both are great—the payoff is (100, 100).

However, if one enterprise acts in concert, the other enterprise does not, then the enterprise that acts in concert will have the least payoff, only 10, whereas the enterprise that does not act in concert will have a higher payoff, up to 80. In Table 1.13, (A acts in concert, B does not) and (A does not act in concert, B does) are both of this kind of situation.

Table 1.13 An example of risk control

	B acts in concert	B does not act in concert
A acts in concert	100, 100	10, 80
A does not act in concert	80, 10	60, 60

Another situation is when both sides adopt the strategy 'not to act in concert', that is the strategy profile for both sides is (A does not act in concert, B does not act in concert). Here, because both sides are prepared, the loss for both sides is smaller than if one side acts in concert but the other side does not. Both sides can obtain 60 units of payoff.

Underlining the optimal strategy, it can be seen that there are two pure strategy Nash Equilibria in this game; they are (A acts in concert, B acts in concert), and (A does not act in concert, B does not act in concert).

In the Nash Equilibrium point (A acts in concert, B acts in concert), the payoff for both sides is (100, 100); in the Nash Equilibrium point (A does not act in concert, B does not act in concert), the payoff for both sides is (60, 60).

If the two sides can consult, it is very likely that they will choose (A acts in concert, B acts in concert), so that both sides can obtain the biggest payoff.

The problem is that this is a situation of non-cooperative game, that is, decisions are made on the basis of neither side having confidence in the other side. In such situation, from the point of view of preventing risks, (A acts in concert, B acts in concert) is not likely to be the game's outcome.

For example, A will consider: If I choose to 'act in concert', but the other side, for some reason (such as misunderstanding, or being provoked by a third party, or lack of resources within the enterprise, etc.) chooses to 'not act in concert', then my own payoff will reduce sharply from the originally expected '100' to '10'. If I choose to 'not act in concert', then in the worst case scenario, I am guaranteed '60' units of payoffs.

Similarly, B will also consider this problem.

Therefore, to control risks in this game, the most likely outcome is (A does not act in concert, B does not act in concert).

1.7.2 The Principle of Pareto Optimum

The principle of Pareto optimum means the measure by which the Nash Equilibrium is chosen by the various players, considering only the payoffs.

Where there are many Nash Equilibria, if on one of the Nash Equilibrium, the payoff for each player is bigger than one's own on other Nash Equilibrium. Here, if each player only considers payoff, then this Nash Equilibrium will be chosen unanimously as the outcome of the game.

Still using the game example in Table 1.13, clearly for the two Nash Equilibria points (A acts in concert, B acts in concert) and (A does not act in concert, B does not act in concert), the Pareto optimal Nash Equilibrium is (A acts in concert, B acts in concert).

1.8   Nash Equilibrium in Dynamic Games

Seeking Nash Equilibrium in dynamic games is mainly via the rollback method, that is, assuming that each player can observe and judge accurately his own payoff and that for other players from the various strategy (action) portfolios, then starting from the payoff at the end point of the game tree to make a judgement on the payoff for the various players. From this a guess is made on the action chosen by the player before this outcome was achieved, an action by which the player was seeking a maximum payoff.

Below, we still use the husband and wife game where the female makes the first move (Fig. 1.1) as an example, to explain the rollback process.

The specific process in a rollback:

Firstly if the wife has already chosen the 'husband's home', let us see how the husband will choose. From the top half of Fig. 1.1, it can be seen that the wife has already chosen the 'husband's home'. If the husband then chooses the 'husband's home', then the payoff for the husband is '2', and the wife gets '1'. However, if the husband chooses the 'wife's home', then the outcome is the husband goes to the wife's home to spend the Chinese New Year, and the wife goes to the husband's home to spend the Chinese New Year. Neither side can go to their own parents' house for the Chinese New Year, nor can they be together. Therefore both the husband and wife get '−1'. Thus, it can be seen that if the wife has already chosen to go to the husband's home for the Chinese New Year, then considering the payoff, the husband will definitely choose the 'husband's home'. Now, having chosen the 'husband's home', the wife's estimate of her own payoff is '1'.

If the wife has already chosen the 'wife's home', let us see how the husband will choose. From the bottom half of Fig. 1.1, it can be seen that the wife has already chosen the 'wife's home'. If the husband chooses the 'wife's home', the outcome is

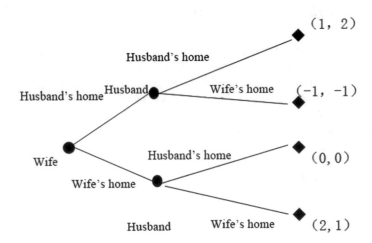

Fig. 1.1   The wife makes the first move in a husband and wife game

that husband and wife will spend the Chinese New Year with the wife's family, then the payoff for the husband is '1', and for the wife is '2'. But if the husband persists in choosing the 'husband's home', then the outcome is that the husband goes back to his home for the festival, and the wife goes back to her home; both sides can only go to their own parents' house for the Chinese New Year but cannot be together. Therefore, the payoff offsets the loss, and both sides get '0'. Thus, it can be seen that if the wife has already chosen to spend the Chinese New Year at her home, then from the point of view of his own payoff, the husband will also choose the 'wife's home'. Thus, for the wife, she can evaluate that, having chosen the 'wife's home', her own payoff is '2'.

To summarise the above situation: if the wife makes the first move, then she can guess that if she chooses the 'husband's home', her own payoff will be '1'; if she chooses the 'wife's home', then her own payoff will be '2'. Therefore, after a payoff evaluation, the wife will choose the 'wife's home'.

This is the dynamic game Nash Equilibrium reached by rollback: The wife makes the first move and she chooses the 'wife's home'; the husband makes a later move, and he chooses the 'wife's home'.

Note that generally the dynamic game Nash Equilibrium is connected with the order of the actions of the various players. Later in this book, it will be seen that if the husband makes the first move, then the outcome for the Nash Equilibrium of this game is the husband chooses the 'husband's home', and the wife chooses the 'husband's home'.

Chapter 2
Low Efficiency Caused
by Non-cooperative Games

In non-cooperative games each player chooses the action which is most beneficial to themselves (this is individual rationality), in the end achieving Nash Equilibrium. Such action under individual rationality is often not as much as the payoff for each player brought by collective rationality (action as guided by the payoff for all players). That is, under some circumstances, individual rationality is often inconsistent with collective rationality, and it destroys the action chosen under collective rationality.

2.1 Prisoner's Dilemma

The Prisoner's Dilemma is a classic case in games. It tells of the police arresting two suspects A and B but without sufficient evidence to prosecute the two. Then the police keep the suspects in separate cells, interrogate the two separately and give them the following choices.

If you confess and plead guilty, whereas the other denies, then you will be released immediately, and the other is sentenced to 10 years imprisonment.

If both deny, then both will be sentenced to 1 year imprisonment due to insufficient evidence.

If both confess, their crime is confirmed, but because of their confession they will both be sentenced to 8 years imprisonment (Table 2.1).

Table 2.1 Prisoner's dilemma

	B denies	B confesses
A denies	Serve a 1-year sentence, serve a 1-year sentence	Serve a 10-year sentence, immediately released
A confesses	Immediately released, serve a 10-year sentence	Serve an 8-year sentence, serve an 8-year sentence

© China Economic Publishing House and Springer Nature Singapore Pte Ltd. 2018
S. Sun and N. Sun, *Management Game Theory*,
https://doi.org/10.1007/978-981-13-1062-1_2

Thus, there are two actions for each prisoner: Deny or confess.

The character of this game is that no matter what the other chooses, the optimal choice for each is always to confess: If the other denies and I choose to confess, I will be released; choosing to deny I will be given a one year sentence; if the other confesses and I choose to confess, I will be sentenced to 8 years; choosing to deny I will be given a 10-year sentence.

Let us suppose that both suspects understand game theory and understand the reasoning, and thus will both choose to confess. The outcome is that each will be sentenced to 8 years.

On the contrary, if they cooperate, for instance they are trusted friends, and have agreed beforehand that if arrested, they will choose to deny. If both sides abide by their promise, then both will only serve a one-year sentence.

It can be seen from this classic game that non-cooperative games lead to low efficiency.

2.2 Competition and Cooperation for Duopoly—The Cournot Model and Cooperation Payoff

The Cournot model (this example is taken from *Game Theory and Information Economics* by Zhang Weiying) describes the process of how two manufacturers who produce a homogeneous product reach equilibrium for their output decisions in the game. The conditions are:

A certain product is manufactured by two enterprises 1 and 2; the production cost of that product is zero.

The function of the product's market price is $P = a - (q_1 + q_2)$. Within this, P is the product's selling price, a is the constant, and q_1 and q_2 are the output of enterprises 1 and 2. The two enterprises also accurately understand the demand curve in that market.

Under non-cooperative game conditions, the two enterprises will first estimate the other's output, and then determine their own output that will give themselves the biggest profit.

More importantly, when the two enterprises decide on their own output, it must be done simultaneously without the situation of one enterprise determining its output before the other.

Let us suppose that:

q_1 is enterprise 1's output, q_2 is enterprise 2's output, u_1 is enterprise 1's profit, and u_2 is enterprise 2's profit, then:

$$u_1 = q_1 P = q_1[a - (q_1 + q_2)]$$
$$u_2 = q_2 P = q_2[a - (q_1 + q_2)]$$

Seeking the optimal output for enterprise 1 and 2:

$$\frac{\partial u_1}{\partial q_1} = a - (q_1 + q_2) - q_1 = a - q_2 - 2q_1 = 0$$

$$\frac{\partial u_2}{\partial q_2} = a - (q_1 + q_2) - q_2 = a - q_1 - 2q_2 = 0$$

Solving the optimal output for enterprise 1 and 2:

$$q_1^* = \frac{1}{2}(a - q_2)$$

$$q_2^* = \frac{1}{2}(a - q_1)$$

Everything being equal between the two enterprises, solving the above simultaneous equations will arrive at the Nash Equilibrium of the simultaneous decisions on output by the two enterprises:

$$q_1{}^* = q_2{}^* = \frac{a}{3}$$

Here, the payoff for each enterprise is:

$$u_1 = q_1 P = q_1[a - (q_1 + q_2)]$$
$$= \frac{a}{3}\left[a - \left(\frac{a}{3} + \frac{a}{3}\right)\right] = \frac{a}{3} \times \frac{a}{3}$$
$$= \frac{a^2}{9}$$
$$u_2 = \frac{a^2}{9}$$

Sun Shaorong re-modelled this classic duopoly non-cooperative game into a duopoly cooperative one. Let us suppose that the two enterprises reach a cooperation agreement. The method for setting output is to first determine the reasonable total output together by both enterprises. Then this total output is divided by 2, which becomes the output for each enterprise.

As such, let us suppose that the total output for the two enterprises is q, and the total payoff for the two enterprises is u.

Then the group's payoff is:

$$u = qP = q[a - q]$$

Seeking the group's biggest payoff, to take the derivative of the overall output and make it 0, we have:

$$\frac{\partial u}{\partial q} = [a - q] - q = 0$$

The solution is:

$$q^* = \frac{a}{2}$$

Here, since the two manufacturers are homogenous and thus under the same conditions, the output for each is half of the total output:

$$q_1{}^* = q_2{}^* = \frac{q^*}{2} = \frac{a}{4}$$

Here, the payoff for each enterprise is:

$$u_1 = q_1 P = q_1[a - (q_1 + q_2)]$$
$$= \frac{a}{4}\left[a - \left(\frac{a}{4} + \frac{a}{4}\right)\right] = \frac{a}{4} \times \frac{a}{2}$$
$$= \frac{a^2}{8}$$
$$u_2 = \frac{a^2}{8}$$

It can be seen that under the conditions put forward by Sun Shaorong in the duopoly cooperation model, the payoff for both enterprises is higher than that of Cournot's model of non-cooperative game.

At present, China faces the problem of massive overcapacity. For example, there are large surpluses in the production of steel, coal and construction materials etc., which also lead to low efficiency for the enterprises in these industries. Such serious overcapacity can be explained in light of the game theory: the local enterprises make decisions independently rather than reach a unified decision cooperatively. This is the same reasoning as to why the output in a competitive duopoly is greater than that in a cooperative situation.

2.3 The Irrational Equilibrium in the Tragedy of the Commons and Sun Shaorong's Fishing Model, and How the Number of People in a Collective Affects Efficiency

2.3.1 An Introduction to the Tragedy of the Commons and Sun Shaorong's Fishing Model

The Tragedy of the Commons is a well-known case of non-cooperative game by multiple entities. This game is designed for institution problems by Hardin. Under institution of publicly owned production resources, and production payoff going to individuals, equilibrium is achieved by non-cooperative games of the many entities concerned, which leads to the depletion of publicly owned resources and reduces the payoff for everybody. Hardin has proved that the payoff for each individual can be much higher than in non-cooperative games, if the many entities are united under a leader who coordinates the production actions for everybody in their interests. This in fact is also the problem of low efficiency in non-cooperative games.

In Hardin's model, each herdsman's number of sheep grazing in the field is the variable of the action in the game. The basic idea is that if each herdsman has a large number of sheep, then the total number of sheep in the field is very large, leading to the field being inadequate in supplying the needs for such large number of sheep, which then causes a drop in the value of each sheep (such as the sheep gets thinner). Of course too few sheep can also lead to reduced payoff. Therefore, the problem facing each herdsman is how many sheep he should keep in order to maximise his own payoff.

In Hardin's model, the direct object for consideration is the 'value of the sheep', which is indirectly related to the problem of depletion of the production resource (the field) as a result of a system of publicly owned production resources but with payoffs going to individuals. For this, Sun Shaorong has re-designed a 'fishing model' for international waters (Sun Shaorong, Journal of Systems & Administration, 2008 issue 2). This model reflects a 'natural reality', which is that all fishery resources in international waters has no owner (publicly owned), but the catch each time goes to the fishermen. In this model, the main variable is the frequency of fishing of each fisherman. In a non-cooperative game there are too many individuals (fishermen) carrying out unrestrained fishing at high frequencies, leading to a reduction of the fishery resources, which causes a reduced catch each time for the fishermen, and which leads to the problem of low efficiency in a non-cooperative game between the fishermen.

Furthermore, Sun Shaorong demonstrated how efficiency is affected by the number of fishermen in non-cooperative games. It is pointed out that in a non-cooperative game situation, the more the fishermen, the lower the efficiency. Furthermore from this, the limit value for efficiency loss, or irrational equilibrium, is reached when there is an unlimited number of fishermen in a non-cooperative game. At the same time, the importance of taxation in managing low efficiency caused by non-cooperative games is pointed out. Taxation can lower the frequency of fishing by the fishermen, thus acting to protect the resources and indirectly raising the fishermen's payoff; taxation does not just redistribute income as said in economics books.

2.3.2 Sun Shaorong's Fishing Model

Let us suppose that there are n fishermen fishing in public fishery. The ability of each fisherman is the same. The payoff each time for each fisherman is G. It is proportional

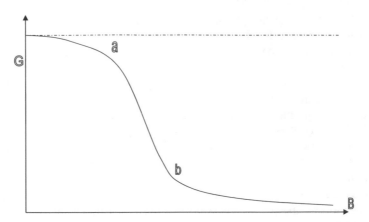

Fig. 2.1 Relationship curve of payoff G per person per catch and frequency of fishing B

to the fish stocks at the time. B is the number of times all fishermen go fishing (i.e. frequency of fishing) per unit of time (such as a month).

Since the payoff G for each fisherman's catch each time is proportional to the fish stocks in the sea, while the fish stock in the sea decreases with the number of fisherman for each unit of time B (i.e. frequency of fishing) in that sea area. Therefore the payoff G for each fisherman will reduce each time with the frequency of fishing B. However, this reduction is not linear.

When B is very small, there is a lot of fish in the sea, and so reproduction is strong. When B starts to increase, the speed of decrease of fish in the sea is not very fast (the left-hand portion in Fig. 2.1).

As B continues to increase, there is less and less fish in the sea, that is, the ability of the fish to replenish—its ability to reproduce is much reduced compared to earlier times. Therefore, the fish stocks in the sea decrease quickly (the portion between *a* and *b* in Fig. 2.1).

In the end, when there is very little fish in the sea, the fishermen's catch gets very little each time. Therefore the decrease of fish stocks at this time becomes fairly slow as the number of fishing increases within each time unit (the right-hand portion in Fig. 2.1 *b*).

In summary, the relationship between payoff G per fisherman per catch and the frequency of fishing B is a curve as shown in Fig. 2.1.

From Fig. 2.1 it can be seen that G is the decreasing function of B. Thus, the following relationship exists:

$$\frac{dG}{dB} < 0$$

For this fishing problem, there is a game relationship between n fishermen. The action of each fisherman is to determine his own number of fishing within each time unit given n fishermen are fishing at the same time, that is one's own frequency of fishing to maximise one's payoff.

Let us suppose that when the n fishermen's game reaches Nash Equilibrium, the frequency of fishing is B^*, then:

$$B^* = \sum_{i=1}^{n} b_i$$

In this formula, b_i indicates fisherman i's number of fishing in the unit of time, i.e. fisherman i's frequency of fishing.

When a fisherman's frequency of fishing is b_i, the payoff is:

$$u_i = G(B)b_i - cb_i \tag{2.1}$$

Of this, $G(B)$ is the earning for each fisherman in a single catch (i.e. payoff before cost deduction); it is the decreasing function of B which is the frequency of fishing in that sea area; c is the production cost per person per catch.

In order to seek the optimal frequency of fishing b_i^* of each fisherman, we take the derivative of b_i

$$u_i' = G'(B)b_i + G(B) - c$$

When fisherman i's frequency of fishing is the optimum value b_i^*, he achieves the biggest payoff u_i^*. Here $u_i' = 0$, that is

$$G(B) + b_i^* G'(B) = c \tag{2.2}$$

When every fisherman achieves his optimal frequency of fishing, the total frequency of fishing is B*, that is $B^* = [b_1^*, b_2^*, \cdots, b_i^*, \cdots b_n^*]$. Here no fisherman will unilaterally change his own frequency of fishing, therefore $B^* = [b_1^*, b_2^*, \cdots, b_i^*, \cdots b_n^*]$ is the Nash Equilibrium point.

The process of that game is completely symmetrical, and the condition of each fisherman is the same, therefore

$$b_1^* = b_2^* = \cdots = b_i^* = \cdots = b_n^* = \frac{B^*}{n}$$

Therefore, to satisfy the conditions for the total frequency of fishing B^* under Nash Equilibrium (i.e. to seek a formula to solve B^*) is:

$$G(B^*) + \frac{B^*}{n} G'(B^*) = c \tag{2.3}$$

In Formula (2.3), n is the total number of fishermen, c is the production cost per fisherman per catch. $G(B^*)$ is the increase of a fisherman's payoff brought by an increase to fish each time in unit time when the frequency of fishing for all the fishermen as a whole is B*. $\frac{B^*}{n}$ is the frequency of fishing for each fisherman when the total frequency of fishing is B*. $\frac{B^*}{n} G'(B^*)$ is when the total frequency of fishing is B*, if each fisherman decides to increase fishing by once per unit of time, the total value of his reduced payoff resulting from it for himself in that unit of time for all the $\frac{B^*}{n}$ of fishing ($G'(B^*) < 0$).

The physical significance of (2.3) is:

As every fisherman knows the payoff $G(B)$ for each catch and the total frequency of fishing B of all the fishermen is of an inverse relationship, let us suppose that in that group (assuming that there is only one group in the whole international sea) the frequency of fishing for each fisherman is $\frac{B}{n}$. Each fisherman decides on his own frequency of fishing (i.e. decide on the value of $\frac{B}{n}$). When $G(B)$, the increased payoff for a fisherman brought by one fishing added by himself in each unit of time, is equal to the sum of $\frac{B}{n} G'(B)$, his own reduced payoff within that unit of time (each fisherman will fish $\frac{B}{n}$ times per unit of time) resulted from that increased fishing (causes increase in frequency of fishing, which in turn causes decrease in payoff each time), and c, increased production cost caused by the additional fishing, increased frequency of fishing individually will not increase payoff. It is at this time that each fisherman will decide not to increase his own frequency of fishing. By this time, B is stable on a certain value of B*, so that the quantity of fish will also be stable at certain level. Whereas the frequency of fishing of each fisherman is stable on $b_i{}^* = \frac{B^*}{n}$.

$b_i{}^*$ is a derivative of u_i from b_i, that is to seek optimal frequency of fishing from the point of view of individual fishermen, therefore what is achieved here is the equilibrium of frequency of fishing B^* which is the Nash Equilibrium point reached with individual rationality.

2.3.3 Sun Shaorong's Non-rational Behaviour Equilibrium Point $B_0{}^*$

The problem is, for a real game with many fishermen, the fishermen in fact cannot know the precise relationship between frequency of fishing and payoff (i.e. the 2.1 formula). Therefore it is difficult to determine for oneself the optimal frequency of fishing. Under most circumstances for the fishermen, an individual fisherman actually does not know how reduced fishing resources relate to his own fishing behaviour, but what he can see is the payoff and production cost for himself brought by one additional fishing.

Therefore the frequency of fishing as judged and determined by each fisherman himself generally will not reach stability at the Nash Equilibrium point as given in Formula (2.3). If, according to economics theory, people will not enter a sector only when payoff equals cost in a competitive situation, then fishermen will only really stop increasing the frequency of fishing when the fishery resources are reduced to such an extent that the payoff for each fishing equals the cost. That is

$$G(B_0{}^*) = c \qquad (2.4)$$

Compared with (2.3) formula, in (2.4) formula the negative item $\frac{B^*}{n}G'(B^*)$ is missing on the left, whereas $G(B)$ is a decreasing function, so with the increase in frequency of fishing B, the payoff $G(B)$ for each catch will decrease. Therefore, where C is the same, $B_0{}^*$ will be greater than the Nash Equilibrium point B^* in Formula (2.3).

That is

$$B_0{}^* > B^* \qquad (2.5)$$

Therefore, for a non-cooperative, competitive fishing game with many fishermen, the fishermen's actual frequency of fishing is much greater than the optimal frequency of fishing B^* as inferred by individual rationality. In fact, B^* is an idealised game outcome by individual rational judgement and when information is complete, while in Formula (2.4), $B_0{}^*$ is a game outcome not of an ideal situation, but from incomplete judgement or information.

From Formula (2.4) the equilibrium point $B_0{}^*$ is obtained independently by Professor Sun Shaorong; it was not obtained from Hardin's Tragedy of the Commons model. Professor Sun Shaorong calls $B_0{}^*$ the irrational equilibrium point, so as to differentiate it from Hardin's individual rationality equilibrium point B^*. The significance of this is that $B_0{}^*$ is different from Nash Equilibrium, being a new equilibrium point.

2.3.4 Collective Rationality Equilibrium Point B**

If n fishermen are seen as a collective, to maximise benefits, the optimal first-order condition is:

$$G(B^{**}) + B^{**}G'(B^{**}) = c \qquad (2.6)$$

In the formula c is the cost per person per catch when all the fishermen within a unit of time carry out fishing, that is when B^{**} is the number of the total actions of all fishermen. When the intensity of the action is B^{**}, for each additional catch in each unit of time, the increase in the number of catches in the unit of time leads to reduced payoff for each catch, which leads to a reduced total value $B^{**}G'(B^*)$ for B^{**} catches. $G(B^{**})$ is the increased payoff per catch when the intensity of the action is B^{**}.

Therefore, the physical significance of Formula (2.6) is:

The fishermen's collective is a unified decision-making body. This body knows the relationship between the payoff $G(B)$ per catch and the total number of fishing B by all the fishermen, which is the mathematical model (2.1). The fishermen's collective decide on their frequency of fishing in the unit of time (i.e. decide on the value of B). When $G(B)$, the increased payoff for the collective brought by one fishing added in each unit of time, is equal to the sum of $BG'(B)$, the collective's reduced payoff within that unit of time (the collective will fish B times per unit of time) resulted from that increased fishing (causes increase in total times of fishing, which in turn causes decrease in payoff each time), and c, increased production cost caused by the additional fishing, the increased number of fishing by now will not increase payoff for the collective. The collective will then decide not to increase the number of fishing. By this time, B is stable on a certain value of B^{**}, so that the quantity of fish will also be stable at a certain level.

B^{**} is called the collective rationality equilibrium point, abbreviated as collective equilibrium.

2.3.5 Comparisons of Non-rational Equilibrium Point $B_0{}^*$, Individual Rationality Nash Equilibrium Point B^*, and Collective Rationality Equilibrium Point B^{**}

According to Formula (2.5), we already have:

$$B_0{}^* > B^*$$

Therefore comparing B^* and B^{**} will give us the relationship between the three. Observing conditions for Nash Equilibrium (2.3)

$$G(B^*) + \frac{B^*}{n} G'(B^{**}) = c.$$

And conditions for a collective optimal point of equilibrium (2.6)

$$G(B^{**}) + B^{**} G'(B^{**}) = c$$

For an individual, the condition for equilibrium is:

$$G(B) - c = \frac{B}{n} G'(B)$$

The payoff from an additional fishing minus the cost of that fishing, equals the sum of reduced catch volume each time for the individual in the unit of time resulted from that additional fishing.

For the collective, the condition for equilibrium is:

$$G(B) - c = BG'(B)$$

The payoff from an additional fishing minus the cost of that fishing, equals the sum of reduced catch volume each time for the collective in the unit of time resulted from that additional fishing.

Given that additional number of fishing in a unit of time leads to reduced payoff for each fishing, since the number of fishing for a collective in a unit of time is bigger than that for an individual, with the same frequency of fishing B, for each additional number of fishing, the scale of payoff reduction is bigger for the collective than it is for the individual.

In other words, the payoff for an additional number of fishing allowable at equilibrium as a decision by a unified collective $G(B^{**})$ must be bigger than that decided by an individual $G(B^*)$.

That is $G(B^{**}) > B^*G(B^*)$.

At the same time, $G'(B) < 0$ therefore $G(B)$ is a decreasing function, which gives:

$$B^{**} < B^*$$

Combining the results (2.5), we have:

$$B^{**} < B^* < B_0^* \tag{2.7}$$

That is to say, when n as the number of individuals that compose a collective is large enough:

Collective rationality equilibrium point B^{**}, which maximises the payoff for the whole collective, is the minimum, that is the smallest number of fishing. If at this point equilibrium is achieved (i.e. the number of fishing does not increase), the highest payoff for each member can be achieved.

Based on the fact that individual rational Nash Equilibrium B^* is bigger than the collective benefits equilibrium point B^{**}, if equilibrium is achieved at this point, (i.c. the number of fishing does not increase), the payoff for each member is not as high as the payoff for the collective at the optimal equilibrium point B^{**}.

Since an individual's judgement arrived at instinctively not to increase the number of fishing in a unit of time, i.e. irrational equilibrium B_0^* is bigger than Nash Equilibrium point B^*, it is much bigger than the number of fishing in a unit of time B^{**} that maximises collective benefits. In other words, if there is no interference by a manager to represent the interests of the collective, the intensity of action for equilibrium achieved by individual instinctive judgement far exceeds that of collective rationality equilibrium point.

2.3.6 True Equilibrium Point—Depending on Society's Average Profit

If we take into account the reality of the flow of capital and labour that exists between the different sectors in society, then whatever the sector (including fishing) is, people will only stop entering the sector when the payoff minus cost (i.e. net payoff) in that sector equals society's average profit. Therefore, fishermen will truly stop increasing the frequency of fishing when the fishery resources are reduced to the extent that the payoff minus the cost for each action in fishing leaves a net payoff that equals society's average profit.

$$G(B_{00}{}^*) - c = K \qquad (2.8)$$

In the above formula, K is society's average profit, $B_{00}{}^*$ is the equilibrium point at which the payoff from fishing equals society's average profit.

Thus, there are four possible equilibrium points in the fishing problem. As society's average profit differs, there are two scenarios in the possible relationship between them:

$$B^{**} < B^* < B_{00}{}^* < B_0^* \qquad (2.9)$$

$$B^{**} < B_{00}{}^* < B^* < B_0^* \qquad (2.10)$$

Of these, Formula (2.10) is a scenario where society's average profit is relatively high, whereas Formula (2.9) is a scenario where society's average profit is relatively low.

2.3.7 The Influence of the Quantity of Collective (n) on Equilibrium Point

Now let us consider the effect the number of members in a collective has on individual rationality equilibrium point B^*.

For this, we re-examine the Nash Equilibrium formula of individual games.

$$G(B^*) + \frac{B^*}{n} G'(B^*) = c \qquad (2.3)$$

Regarding the effect the number of members n in a collective has on the individual rationality equilibrium point, the conclusion mainly is as follows.

2.3.7.1 The Bigger n Is, the Further Nash Equilibrium Leans Towards Over Equilibrium

Looking at the Nash Equilibrium Formula (2.3), it can be seen that if there are many members in the collective, i.e. as n gets towards infinity, Formula (2.3) becomes Formula (2.4):

$$G(B_0^*) = c \tag{2.4}$$

Namely the formula for Nash Equilibrium of individual rationality becomes the individual irrational equilibrium formula. That is, as the number of members increases in a collective, even if individuals can make judgement based on individual rationality, the equilibrium point will still lean towards individual irrational excessive action. This means that for collectives without the restraint of managers, the more the members, the more serious the depletion and damage to resources.

2.3.7.2 The Smaller n Is, the Further Nash Equilibrium Leans Towards Collective Rationality Equilibrium

Looking at the Nash Equilibrium Formula (2.3) for individual games, it can be seen that if n becomes small, Nash Equilibrium based on individual rationality will change towards collective rationality equilibrium. In particular, when n equals 1, the original Nash Equilibrium completely becomes collective rationality equilibrium.

$$G(B^{**}) + B^{**}G'(B^{**}) = c \tag{2.6}$$

That is, for collectives without the restraint of managers, the fewer the members, the less severe the depletion and damage to resources. This suggests that if the resources can be divided, then the equilibrium in a system of privately owned resources with profit-seeking behaviour by many members is the same as that in a system with publicly owned resources to maximise collective interests.

Chapter 3
Free Ride, Adverse Selection, Moral Hazard and Separating Equilibrium

3.1 Free Ride Under Conditions of Symmetrical Information

A free rider refers to someone who enjoys others' outcome without exerting efforts. To exert efforts implies a cost is involved, so to free ride implies there is no cost to oneself for hard work but a free rider can enjoy the fruits of others who paid the cost for hard work. Therefore to free ride is actually a behaviour that takes advantage of others. Free ride is a classic game behaviour.

The Theory of free ride comes from the book *The Logic of Collective Action: Public Goods and the Theory of Groups* (published in 1965) by the economist Mancur Olson.

Example 3.1 The boxed pigs game

The 'boxed pigs' game is a classic case in game theory put forward by John Nash in 1950. The content of the game is: There is one big pig and one small pig in a pigsty. At one end of the rectangular pigsty is a trough, and at the other end is a switch that controls the feed in the trough. Press the switch and 10 units of pig feed will get in the trough.

It takes a certain amount of time to return to the trough from where the switch is, therefore:

If the big pig goes to press the switch, the small pig waits by the trough, then the small pig, which eats slowly but ahead of the big pig, can eat 4 units of feed. The big pig can eat 6 units of feed.

If the two pigs go together to press the switch, then they both reach the trough at the same time. The result is that as the big pig eats fast, it can eat 7 units of feed, whereas the small pig can only have 3 units of feed.

If the small pig goes to press the switch, the big pig waits by the trough, then the big pig can eat up all 10 units of feed. By the time the small pig reaches the trough, there is no pig feed left.

Of course, if neither the big pig nor the small pig goes to press the switch but wait by the trough, then there is not any feed in the trough, so that the feed for both small pig and big pig is 0 (Table 3.1).

In this game the strong Nash Equilibrium is the big pig pressing the switch while the small pig waiting by the trough. Here, the big pig can eat 6 units of feed whereas the small pig can eat 4 units.

The situation of the boxed pigs game achieving equilibrium point (big pig presses switch, small pig waits) is that of a small enterprise taking a free ride, because without contributing labour it enjoys the fruits from the big enterprise.

The boxed pigs game is common in business sectors. For instance it is usually small enterprises that wait for big enterprises to develop the technologies and market. When the market matures, the strategy of small enterprises is to follow and imitate.

In product sales many enterprises let weak products follow strong products to leverage for 'distribution'. In fact this is a kind of free riding behaviour. 'Magic yak bone marrow granules' free riding with 'Biyang bone strengthening yak bone marrow powder' is an example. 'Biyang bone strengthening yak bone marrow powder' has bombarded television and newspaper media with intensive advertising. In the end, 'Magic yak' adopts the strategy to follow 'Biyang' closely: where there is 'Biyang yak', there is 'Magic yak', and 'Magic yak' thus got very good sales performance. In appearance the packaging of 'Magic yak' is almost the same as 'Biyang yak', but is slightly cheaper.

In management, 'free riding' is often harmful—it weakens the motive for hard work.

For instance, China experienced a period of equal distribution, that is, the salary for each member of staff in an enterprise was fixed—no more for working harder and no less for working less hard. The result was to seriously dampen the enthusiasm of the hard workers and encouraged the lazy. And over time, the hard workers also became lazy.

Table 3.1 Boxed pigs game payoff matrix

		Small pig	
		Press switch	wait
Big pig	Press switch	7, 3	6, 4
	Wait	10, 0	0, 0

3.2 Asymmetric Information and Adverse Selection

3.2.1 Asymmetric Information

Asymmetric information in a game means the information possessed by the two parties is different. In real life asymmetric information is a very common phenomenon. In such situation, the party with more information is often at an advantage.

For instance during trading, buyers and sellers often possess different information. In some situations, the seller has more information than the buyer with regards to the goods to be traded, for example in selling used cars, the seller knows the car's performance best. In some other situations, the buyer has more information than the seller, for example in the medical insurance market, the buyer's understanding of his own health conditions is better.

If there is asymmetric information in trading games, it can cause adverse selection before the buyer and seller sign the contract (or sign the agreement), and moral risk after the buyer and seller have signed the contract.

3.2.2 Adverse Selection

In competitive games, if information is complete and symmetrical, the result in competitive games should be 'survival of the fittest'. In trading, good quality wins whereas poor quality goods will be eliminated from the market. However in situations of asymmetric information, because the party with poor information cannot accurately judge the effect of his own actions in the game (such as whether to buy product A or product B), the situation of 'elimination of the best and survival of the poor' often happens in competitive games which is contrary to reason.

Asymmetric information causing adverse selection makes many basic economic principles ineffectual. Apart from the 'survival of the fittest' principle in competitive markets, law of price will also become invalid. According to the law of price, if the price of certain product is lowered, that product's market demand will increase; by contrast, if that product's price is raised, its demand will decrease. However, as consumers have little product information, when its price is lowered, consumers are reluctant to buy that product because they worry about its quality. By contrast, in raising the product price, consumers may be willing to buy because they may believe that the product is of high quality.

Kenneth Joseph Arrow in 1963 began to study asymmetric information that leads to the phenomenon of adverse selection. Later, in 1970 George Akerlof published the thesis 'The Market for Lemons: Quality Uncertainty and the Market Mechanism' which further studied this, and in 2001 was awarded the Nobel Prize in economics.

George Akerlof's theory on adverse selection is called the 'lemons market model'. From the perspective of trading in 'lemon cars' as an example, he explained the problem of adverse selection. Lemon is a fruit that appears nice but is bitter inside, so in American slang 'lemons' are goods with defective quality that cannot be noticed from the outside.

George Akerlof wrote that in the market for used cars, a seller generally knows the quality of the car, while a buyer generally does not know but he knows the average quality in the used car market. In such situation, to avoid risks in all used cars, the buyer, based on his understanding of all used cars of average quality, is only willing to give the seller a medium price (this is game behaviour). As such, sellers of those used cars of better than average quality suffer a loss. Therefore those 'good' cars of quality above average level are withdrawn from the market. The result is that with good cars withdrawn, the average quality drops in the second-hand car market. As a result the price quoted by buyers continues to drop, based on the declining average quality. As this repeats, in the end all the cars in the market are of the 'worst quality'. This process is due to the existence of asymmetric information in games, which gradually push good cars out of the market. But according to the efficient market hypothesis, competition should result in good cars pushing bad cars out of the market. Since this phenomenon is contrary to the conclusion of the efficient market theory, this process is called adverse selection.

The Akerlof model is:

Assume that there are only two types of used cars in the market, one is the high quality cars worth 6000 US dollars, and the other is low quality cars only worth 2000 US dollars.

Buyers do not know the specific quality of each car, but they know that the average quality car in the used car market is 4000 US dollars. Therefore, the highest price the buyer is prepared to pay is 4000 US dollars. Thus, sellers of high quality cars worth 6000 US dollars suffer a loss. Therefore only sellers with low quality cars worth 2000 US dollars are left in the market.

When the buyer knows that high quality cars are withdrawn, he knows that the cars left in the market must be of low quality worth only 2000 US dollars, therefore the price he now quotes is changed to 2000 US dollars. This price is the equilibrium as a result of the game by the buyers and sellers in the used car market. The game leads to the result that only low quality used cars are left in the market; high quality cars have been pushed out. Thus, information symmetry causes 'market failure', that is 'survival of the fittest' has been reversed.

In recent years electronic commerce develops rapidly but the problem of asymmetric information is more acute, leading to an even more serious problem of adverse selection. This is because on the Internet, it is even more difficult to verify product quality, and the vendor's real identity is even more difficult to know.

Asymmetric information that leads to adverse selection also exists widely in the area of human resources. For instance, among company employees adverse selection is frequent. Generally speaking a company manager does not know for sure the ability of each member of staff. In such situation, he pays them an average salary according to the average ability of employees. In such situation, employees who are more able feel that they have lost out, and so choose to leave the company, whereas less able employees feel that it is 'worthwhile' and are willing to stay in the company. Gradually, more able employees are being pushed out of the company, basically leaving behind less able ones.

In Akerlof's model, the seller is the party with superior information. But in the insurance sector, the buyer is often the party with superior information. For instance, in the car insurance market, buyers know more about their own driving habits and probability of claims than the insurance company. Insurance companies cannot differentiate between high risk and low risk customers, and as such they charge each customer the same insurance premium, which is determined by the average probability for accidents. In such situation, only those who are accident prone will actively buy insurance, whereas those drivers who are more careful feel that buying insurance is not worthwhile, and so are not keen to buy insurance. In such situation, most of the customers who buy insurance are high risk people, whereas low risk customers gradually withdraw from the market. This process will gradually lead to a higher probability for claims. For the company, this is adverse selection.

The significance of the theories of asymmetric information and adverse selection is that they reveal certain flaws that exist in the theory of free market. During the times of Adam Smith, the 'invisible hand' of the market was much admired. Mainstream economic theory advocated a self-regulated free market, and opposed any state interference in the market. In reality, asymmetric information is very common in society and it is this kind of situation which hinders social justice and affects the efficiency in the market allocation of resources.

3.2.3 Ways to Solve Adverse Selection

In order to solve the problem of adverse selection that causes market failure, many approaches emerge. The common feature of these approaches is that they are all based on the improvement of information asymmetry.

One is to record and provide evaluation by previous customers. Clearly, customer evaluation is higher for good quality products, or the probability of customers scoring 'good' is higher. This method of recording customer evaluation is widely used in e-commerce. Also, some products that become good quality brands are actually a result of customer evaluations. This information can provide reference for subsequent buyers.

Secondly, manufacturers provide quality assurance, maintenance, returns, etc., to distinguish their own good quality products from the inferior ones.

Thirdly, quality certificates or indicators, etc. can be issued by a third party (such as the government or industry associations). In such situation, the guarantor for quality has shifted from the manufacturer to a more authoritative and trusted third party such as a government, which clearly makes the buyer feel that the quality is more reliable.

In the human resources market, academic and the various qualification certificates awarded by education institutions and examination bodies are an effective approach for improving the problem of asymmetric information in this market. With these certificates, organisations recruiting can evaluate the ability and professional levels of these job applicants. In this respect, according to Michael Spence the more able can obtain qualifications at a lower cost (such as time and effort in studying), whereas those of lower ability must obtain qualifications with a very high cost. Therefore it is more 'economical' to go to university only for those who are highly capable. Thus, academic qualifications are information in the labour market. Organisations can evaluate the applicant's ability using this information, and therefore to improve the problem of asymmetric information in the labour market. According to Spence, certificates of qualifications can be regarded as proactive signals from job applicants to improve the problem of asymmetric information. Therefore his theory is called the 'signalling model'.

Fourthly, information seeking. This mainly refers to activities of the party with inferior information, such as visits, surveys, etc. so as to increase the information they hold. Clearly, this method for improving asymmetric information is rather costly for the individual in the game.

In order to prevent 'adverse selection' companies generally strengthen their performance appraisal for individual staff. Via performance-related pay, employees of high capability get high rewards, and employees of low capability get low rewards. This method is in fact also an improvement for asymmetric information by seeking information, and thus to prevent the phenomenon of adverse selection. The problem is that for certain posts, the individual performance of employees are hard to appraise. The problem of adverse selection in this type of posts is still difficult to solve.

3.3   Separating Equilibrium

Apart from the above four methods for solving the problem of adverse selection, Joseph Stiglitz also proposed the separating equilibrium model. This model is mainly established for the study of asymmetric information in games concerning how the party with weak information differentiates the side with strong information. Joseph Stiglitz mainly studied the phenomenon of asymmetric information in the insurance market and the credit market, and proposed the separating equilibrium model.

The reason for the existence in the insurance market of adverse selection is mainly due to the fact that the insurance company does not know the extent of risk of the policyholders. When the insurance premium is generally of average price, people of low risks feel that buying insurance is not worthwhile, and so withdraw from the insurance market. However, people of high risks are willing to buy insurance as they feel it worthwhile. For the insurance company, bad customers have pushed good customers out of the insurance market. To solve this problem, Joseph Stiglitz suggested that insurance companies should provide different types of contracts for the policyholders to choose. One type is a high excess amount with a low premium; another type is a low excess amount with a high premium. Clearly, if the customer is of low risk, then choosing a contract of high excess with a low premium is more economical. If the customer is of high risk, then choosing a contract of low excess with a high premium is more economical. Thus, having two types of contracts help to differentiate between high and low risk customers, which was originally not possible. This is the separating equilibrium in games.

The separating equilibrium phenomenon also exists when stocks and shares companies issue shares (i.e. IPO).

When companies issue new shares, high quality companies generally have their first tranche issued at a discount, attracting the attention of investors with a fairly low issue price. When the investors know the company better, the company then have a higher price for their seasoned equity offering (SEO). The second issue of shares at a higher price can make up for the cost of the shares issued at a discount the first time. But for an inferior company who fears that investors will know, it will raise as much capital as possible at the first issue. Thus the two different behaviours in issuing shares by a high quality company and an inferior company lead to a separating equilibrium, that is a high quality company raises capital in two stages—the first stage with a big discount for shares issued, whereas an inferior company raises capital in one stage and also issues shares with a small discount.

Example 3.2 Separating equilibrium and pooling equilibrium of job applicants choosing to be educated (get education qualifications) or not (without education qualifications)

Let us assume that there are two types of people in the human resources market—one type is of high ability, the other type is of low ability. The salary institution of the employer will impact on whether it will be a separating equilibrium or pooling equilibrium regarding these two types of job applicants in choosing education (get education qualifications) or not (no education qualifications). That is to say, if the employer has an appropriate salary institution, then in their choice of education, the high ability and the low ability will form a separating equilibrium. However, if the salary institution is inappropriate, then the two will probably form a pooling equilibrium, but the specific circumstances for the pooling equilibrium (i.e. equilibrium at not getting educated or getting educated) depend on the salary institution.

Let us assume that in obtaining the same education qualifications, the high ability applicants and low ability applicants have to pay different costs. Of these the cost is 4 for the high ability ones, and 8 for the low ability ones.

Salary institution 1: The two form a separating equilibrium, i.e. the high ability applicants choose education, and the low ability ones choose no education.

Let us assume that the employer cannot differentiate the actual ability of the job applicants and has to rely on whether they have certificates for evaluating their ability. If they have certificates, they will be given 10 units of salary; otherwise they will be given 5 units of salary. Like this:

The payoff for the high ability applicants with education: $10-4=6*$
The payoff for the high ability applicants without education: $5-0=5$
The payoff for the low ability applicants with education: $10-8=2$
The payoff for the low ability applicants without education: $5-0=5*$
* means the **The Optimal Strategy**.

Under such salary institution, the high ability applicants choose education, and the low ability applicants choose no education, i.e. the game outcome is a separating equilibrium. In such situation, the employer as the party with weak information uses an appropriate salary institution to differentiate between the high ability and low ability applicants.

Salary institution 2: The two form a pooling equilibrium, with both the high ability and low ability applicants choosing education.

If they have certificates, then they will be given 20 units of salary; if they have no certificates they will be 5 units of salary. Like this:

The payoff for the high ability applicants with education: $20-4=16*$
The payoff for the high ability applicants without education: $5-0=5$
The payoff for the low ability applicants with education: $20-8=12*$
The payoff for the low ability applicants without education: $5-0=5$

Under this kind of salary institution, whether they are of high or low ability, they will choose education, i.e. the outcome of the game is a pooling equilibrium.

Salary institution 3: The two form a pooling equilibrium, with both the high ability and low ability applicants choosing no education.

If they have certificates, then they will be given 8 units of salary; if they have no certificates they will be 5 units of salary. Like this:

The payoff for the high ability applicants with education: $8-4=4$
The payoff for the high ability applicants with no education: $5-0=5*$
The payoff for the low ability applicants with education: $8-8=0$
The payoff for the low ability applicants with no education: $5-0=5*$

Under this kind of salary institution, whether they are of high or low ability, they will choose no education, i.e. the outcome of the game is a pooling equilibrium.

Example 3.3 Story of the grey circle—separating equilibrium

During the Song Dynasty in Zhengzhou, a rich man surnamed Ma died. Ma had two wives. Only the second wife gave him a son who was to inherit the family's fortune.

But the first wife colluded with officer Zhao in the county government to abduct the child and expel the second wife from the home. The case got all the way to Kaifeng; Officer Bao presided at the trial. Both the first and second wives said the child is theirs.

Officer Bao did not question further, but ordered some slaked lime and drew a large circle in the court. Officer Bao placed the child in the centre, and let the two women each hold one of the child's arms. He said: Whoever can pull the child out of the circle to her side, the child will be hers.

Officer Bao called out 'Start'. The first wife pulled hard. Seeing that the child's arm, thin and small and was being pulled forcibly by the first wife, the second wife hesitated, and the child was pulled over to the side of the first wife.

When Officer Bao saw this, he said that it did not count as the second wife would not pull hard.

The second time, again the first wife pulled forcibly. To start with, the second wife also pulled hard, but the child started to cry. She could not bear it and let go of her hand. Again the child was pulled over to the side of the first wife.

The second wife burst out crying, and explained the reason for not pulling hard. Everyone in court was moved and convinced.

Officer Bao banged on the desk and demanded the first wife to confess. The mystery got cleared up.

This example shows how, a game devised by Officer Bao between the two wives, formed a separating equilibrium. The first wife was not the child's real mother. The biggest benefit was to pull the child over, and so she chose to use brute force to pull the child. But the second wife was the child's real mother. She loved the child. The biggest benefit was for the child not to be hurt, so she chose to let her hand go. Originally Officer Bao was the party with weak information. He did not know who the child's mother was. But having devised a game of separating equilibrium, he successfully differentiated between the real and false mother.

Example 3.4 Time-based pricing strategy of Airlines—separating equilibrium

There are two types of airline passengers:
One is the business travellers—they are not sensitive to ticket prices but are sensitive to time, and require reasonable departure times. The other are travellers like the students—they have no income and thus are very sensitive to ticket prices but are not sensitive to departure times, and basically have few requirements.

Thus, if the airline only has one flight time, setting ticket prices could be a dilemma: If ticket prices are set high, they will lose the student travellers; if set low, they will lose income.

Therefore the optimal game strategy for an airline is: Different ticket prices are set according to flight schedules, i.e. high ticket pricing for flights departing during office hours, but low ticket pricing for flights departing in the night and at dawn. Thus, a separating equilibrium occurred for student travellers and business travellers.

Example 3.5 Additional incentives undermine the separating equilibrium—a huge increase in the number of 'useless patents'

Lincoln has a famous saying: The exclusive use of his invention; and thereby added the fuel of interest to the fire of genius. Patent is a reward mechanism for promoting technological inventions. Venice which was commercially advanced, was the first to implement a system for patents, and promulgated the patent law in 1474. At present, the system for patents is widely used in many countries around the world.

Patent as a concept in law is basically that for an invention to be authorised as a Patent, the applicant must make it known to the world so that society can understand the applicant's invention and progress in related fields. However, patent inventions cannot be used by others free of charge. If someone uses the invention without permission from the patent holder, it will be regarded as an 'infringement of a patent', and lead to a penalty applied in accordance with the country's patent law.

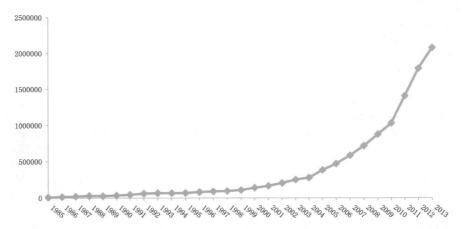

Fig. 3.1 The growth of patents authorised in China. *Source* State Intellectual Property Rights Office
Annual Report (http://www.sipo.gov.cn/tjxx/)

Table 3.2 Some higher education institutions' patent applications and exploitation—survey by Liu
Yue'e et al.

Year	Number of applications	Number of exploited patents	Rate of exploitation (%)
2001	3180	289	9.09
2002	4377	416	9.50
2003	8650	543	6.28
2004	10,940	910	8.32
2005	16,578	976	5.89

Thus, it is necessary to give remuneration to the patent holder to obtain permission
for use, the so-called 'patent purchase'.

In recent years, the number of China's patents is growing unusually fast (see
Fig. 3.1). But at the same time, the rate of exploited patents (the proportion of patents
in actual use as compared with patents applied for) is falling significantly. Table 3.2
is the survey result by Liu Yue'e et al. in 'The current status in the exploitation of
higher education institution patents—an investigation and reflections'. Figure 3.2 is
produced from the results of that survey, showing the status for exploiting of patents
during 2001–2005 in the top 100 schools that applied for patents. It can be clearly seen
from Table 3.2 and Fig. 3.2 that higher education institutions applying for patents
soared from 3180 in 2001 to 16,578 in 2005, but the rate of exploitation dropped
from 9.09% in 2001 to 5.89% in 2005.

The average quality of patents has dropped, and the rate of 'useless patents' as
compared to applications has increased, which is a result of the separating equilibrium
of applying for patents having been undermined. Let us construct the separating
equilibrium model to analyse this situation as follows:

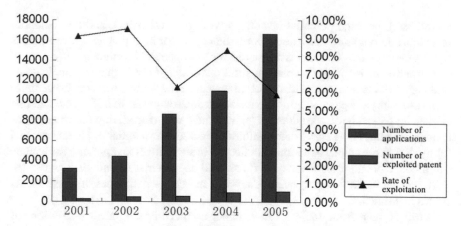

Fig. 3.2 Contrast diagram of patent application and exploitation of higher education institutions plot according to investigation result of Liu Yue'e et al.

Let us assume there are two types of inventions. A type is high potential inventions, i.e. the inventions that can be put into production for use and can produce economic efficiency; B type is of low potential, i.e. the inventions that cannot be put into production for use and so cannot produce economic efficiency.

For any invention (i.e. whether it is A or B), the inventor has two choices for action: to apply for patent or not to apply for patent. Of these, a^+ indicates that the inventor patenting an A type invention; a^- indicates that the inventor not patenting an A type invention; likewise, b^+ indicates that the inventor patenting a B type invention; b^- indicates that the inventor not patenting a B type invention.

Whichever type of invention, the cost in applying for patent (such as application fee and patent annual fee etc.) is $c > 0$. For a high potential invention A, because it can be put into production for use and so produce economic efficiency, therefore after the patent application, the patent assignment cost is $v(a^+) > c > 0$. But for a low potential invention B, because it cannot be put into production for use and so cannot produce economic efficiency, after patent application, the patent assignment cost is $v(b^+) = 0$.

Thus, under a normal patent institution, the inventor's payoff matrix for both types of inventions is as shown in Table 3.3.

It can be seen from Table 3.3 that, under a normal patent institution for high potential inventions A and low potential inventions B, the Nash Equilibrium should

Table 3.3 The inventor's payoff matrix for both types of inventions under a normal patent institution

	b^+	b^-
a^+	$v(a^+) - c, -c$	$v(a) - c, 0$
a^-	$0, -c$	$0, 0$

be (a^+, b^-), i.e. high potential inventions will apply for patent, and low potential inventions do not apply for patent. Here, the payoff for high potential inventions is $u(a^+) = v(a^+) - c > 0$, whereas the payoff for low potential inventions is $u(b^-) = 0$.

Currently the reality is that many institutions aiming to be high up on the 'patents ranking' in the various comparisons often introduce mechanisms to encourage more people to apply for patents. For instance some areas stipulate that if a graduate had applied for patent, he can be allocated 'extra points' when registering for a permanent resident account in the city. Some institutions include the number of patents applied for by staff in the annual assessment indicator. In such situation, applying for patents can produce economic efficiency $v(a^+)$ as well as systemic incentives $s > c > 0$. With such an institution, the inventor's payoff matrix for both types of inventions is shown in Table 3.4.

It can be seen from Table 3.4 that, under an incentivised patent institution for high potential inventions A and low potential inventions B, the Nash Equilibrium should be (a^+, b^+), i.e. high potential inventions will apply for patent, and low potential inventions will apply for patent too. Here, the payoff for high potential inventions is $u(a^+) = v(a^+) - c > 0$, whereas the payoff for low potential inventions is $u(b^+) = s - c > 0$. It can be seen that this kind of incentives undermine the original separating equilibrium, which leads to a kind of pooling equilibrium where inventions of both high potential and low potential apply for patents. In such situation, it is not surprising that the number of useless patents greatly increases.

Likewise, the current criticism on the problem of 'rubbish thesis' is, in effect, due to excessive incentives for theses which undermine the original separating equilibrium.

Also, in many work units everybody is desperate for senior positions and senior job titles, which, due to their many benefits, lead to the original separating equilibrium being undermined. In situations of separating equilibrium, the capable should be the ones seeking senior positions or job titles, while the less capable can then be content doing low position jobs or having low job titles. Now because there are too many benefits attached to senior positions and senior job titles, even though the cost in striving for these is very high for the less capable, once they succeed it is still worthwhile. This is what has caused everybody to strive for senior positions and senior job titles.

Table 3.4 The inventor's payoff matrix for both types of inventions under an incentivised patent institution

	b^+	b^-
a^+	$v(a^+) + s - c, s - c$	$v(a^+) + s - c, 0$
a^-	$0, s - c$	$0, 0$

3.4 Asymmetric Information and Moral Hazard

In modern society, many games are conducted under constraints of bilateral agreements (if the two parties to the agreement is of a principal-agent relationship, then one party is the principal, the other the agent). For instance during working hours a company employee (generally the company is the principal, the employee the agent) is under the constraint of the signed agreement and chooses actions most beneficial to himself. If both parties to the agreement are under conditions of asymmetric information (usually the agent has better information because he knows best whether he is doing his job in accordance with the agreement), one party (usually the agent) uses his advantage in possessing better information to carry out actions meant to be concealed, which are difficult to be observed, so as to profit at the expense of the other party. This kind of behaviour is called moral hazard.

Common moral hazards are:

Being slack, or free riding in situations of asymmetric information, with the individual exerting little while enjoying the fruits of other people's efforts.

Opportunism, or efforts of the individual to increase his own payoff are at the expense of the other party (the principal). This situation is also called negative direction hard work.

There are mainly two conditions that lead to moral hazard. The first is existing conflict on utility between contracting parties. The second is asymmetric information. This kind of asymmetric information is different from that in adverse selection. It emerges after the agreement has been signed, mainly caused by the fact that whether or not one party carries out actions as stipulated in the agreement is difficult to observe. If whether one party carries out actions as stipulated in the agreement is not easily seen by the other party, then the former has a certain advantage over information, and moral hazard can occur.

Example 3.6 Moral hazard in the insurance market

Between the insurance company and the policyholders there exists information asymmetry. The policyholders are with better information because the insurance company cannot observe the actions of the policyholders after signing the insurance contract. For example, after insuring their properties, some of the policyholders do not look after their properties carefully, because when they suffer a loss, they will be compensated by the insurance company. There are even people who deliberately create accidents to defraud insurance premiums. All these behaviours cause insurers to undertake a higher than normal rate for loss.

3.5   Some Approaches for Solving Moral Hazard

In practice, there are mainly three types of approaches for solving moral hazard.

One is to strengthen information observation, so as to minimise information weakness, and reduce information asymmetry.

Secondly, if it is a repeated game, then post penalties can be used, that is if one party in the game exhibits moral hazard, making these behaviours known to the world can create a kind of post penalty, so that moral hazard can be deterred.

The third is to design a reasonable contract, so that the effective goal of the agent is consistent with that of the principal's, thus ultimately preventing moral hazard. In this respect, the economist Hurwiez put forward the 'incentive compatibility principle', that is, by devising a reasonable agreement or system, the individual interest of the agent becomes consistent with that of the principal's. For instance, with regard to sales personnel, a company's remuneration system that is 'in accordance with sales commission' is consistent with the principle of compatible incentives, because under such a system, the sales personnel as the agent will aim for sales quantity, and for the company as the principal, increasing sales is also what they want.

$$\int u(s(r(a^*,\ \theta)))f(r)dr - c(a^*) \geq \int u(s(r(a,\theta)))f(r)dr - c(a) \qquad (3.1)$$

Formula 3.1 is put forward by Hurwiez et al. on 'incentive compatibility principle' expressed as a mathematical model. Here:

a^* is the action the principal hopes the agent will choose, a are other random actions apart from a^* that can be chosen by the agent.

$r(a,\ \theta)$ is the agent's output function, among this, θ is the random factor that affects output. This function shows that the agent's output is affected both by his own choice of action a and the random factor.

$s(r(a,\ \theta))$ is the incentive the principal gives the agent (for example pay or bonus etc.). This function shows that the incentive given by the principal to the agent is influenced by the agent's output r. Normally the two form a positive correlation function, i.e. the higher the agent's output, the bigger the incentive he will get.

$u(s(r(a,\ \theta)))$ is the agent's utility function. This utility function shows that the utility the agent gets from his own action is related to the incentive s given by the principal. This relationship is normally positively correlated, i.e. the more incentives the agent gets, the greater utility (u) he feels.

$f(r)$ this is the distribution density function for output.

$c(a)$ is the cost of the action a.

Formula 3.1 shows that, as the principal, if he wants the agent to choose his desired action a^*, he must let the agent have more effect in taking that action a^* than if the agent chooses other actions a. This is the meaning of 'incentive compatibility'. For instance, if the manager (principal) hopes that the workers (agent) work hard, he must let the hard worker gets a much bigger remuneration than the slack worker.

The reader should note that Hurwiez's 'incentive compatibility' formula is premised on an implied assumption: Let us suppose that utility can be increased and decreased. For instance in Formula (3.1), desired utility $\int u(s(r(a, \theta)))f(r)dr$ is subtracted by cost $c(a)$, which is the agent's overall utility. But in reality, only like payoff can be added or subtracted (for example economic payoff can be added or subtracted by economic payoff), and for an action, its efficiency can only be derived after the total payoff is obtained. For example, someone's business cost is 200,000 yuan, the income is 300,000 yuan. The overall utility cannot be expressed as the utility of 300,000 yuan minus the utility of 200,000 yuan, but as 300,000 yuan minus 200,000 yuan resulting in 100,000 yuan and after this, seek the utility of 100,000 yuan. Therefore, Hurwiez's formula only gives the principle of incentive compatibility, i.e. it gives the conditions that must be fulfilled if 'the agent listens to the principal', but it is not a formula that accurately reflects reality. Thus it cannot be applied directly unless 'utility' is changed to 'payoff', then it can be applied to actual practice.

Currently, some companies devise 'employee stock ownership plan' to try and bind the interests of the staff to that of the company as a whole, so that the staff work hard and voluntarily protect the company's interests, which actually is building 'incentive compatibility'.

Chapter 4
First-Move Advantage and Second-Move Advantage

In a game of two players, if both sides do not take action at the same time, then one player makes a move first, the other player makes a move afterwards. In some situations, who makes the first move and who makes the second can lead to positions of advantage or disadvantage. If the first-mover has advantage, it is called first-move advantage; if the second-mover has the advantage, then it is called second-move advantage.

4.1 First-Move Advantage

4.1.1 First-Move Advantage in Discrete Behaviour

Example 4.1 First-move advantage in a husband and wife game (this example is adapted from *Introduction to Game Theory* by Wang Zeke and Li Jie).

The Battle of the Sexes game is a classic case in game theory. It describes the game process between a man and a woman in choosing whether to watch a football match or to watch a ballet show together at the weekend. Generally, men prefer watching a football game whereas women prefer watching a ballet show. However, the two parties are not willing to be separated to watch football and ballet. Thus the two sides carry on the game under such constraints.

For the game to be 'worth playing' (In fact, whether to watch football or ballet, for lovers deeply in love almost no one is willing to 'play games'. Rather they want to please the other by choosing the programme more preferable to the other side.) and to make that case more appropriate to problems facing the Chinese people, the author has adapted the problem to a young couple working away from home. During the Spring Festival, they face the game problem of whether to go to the husband's home

© China Economic Publishing House and Springer Nature Singapore Pte Ltd. 2018
S. Sun and N. Sun, *Management Game Theory*,
https://doi.org/10.1007/978-981-13-1062-1_4

(husband's home) or to go to the wife's home (wife's home) for the festival. Like the Battle of the Sexes game, the husband wants to go back to his family, while the wife wants to go back to her family, but the husband and wife do not want each other to go separately to their respective family to spend the Chinese New Year (which is the worst outcome for both).

Let us assume that the husband and wife go separately to buy tickets for going home. In such a situation, it is a husband and wife game of first-move advantage, i.e. whoever is first to go to buy tickets will have a higher payoff.

Please take a look at the woman's side making the first move in a husband and wife game (Fig. 4.1).

With inverse method, it can be seen that the equilibrium point of this game is where both the wife and husband choose the wife's home. The outcome is that the payoff for the wife is 2 and the payoff for the husband is 1.

The specific process with inverse method:

Firstly since the wife has already chosen the 'husband's home', let us see how the husband will choose. It can be seen in the top part of Fig. 4.1 that if the husband chooses the 'husband's home', the result is husband and wife both go the husband's home for the Chinese New Year, then the payoff for the husband is '2', the wife '1'. However, if the husband chooses the 'wife's home', then the outcome is the husband goes to the wife's home to spend the Chinese New Year, and the wife goes to the husband's home. Neither side can go to their own parents' house for the festival, nor can they be together. Therefore both the husband and wife get '-1'. Thus, it can be seen that if the wife has already chosen to go to the husband's home for the Chinese New Year, then the husband will definitely choose the 'husband's home'. Thus for the wife, the payoff for choosing the 'husband's home' is '1'.

After the wife has already chosen the 'wife's home', let us see how the husband will choose. It can be seen in the lower part of Fig. 4.1 that if the husband chooses

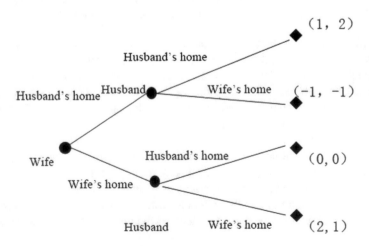

Fig. 4.1 The wife makes the first move in a husband and wife game

the 'wife's home', the result is husband and wife both go the wife's home for the Chinese New Year, then the payoff for the husband is '1', the wife '2'. But if the husband chooses the 'husband's home', then the outcome is that the husband goes home for the Chinese New Year, and the wife goes home for the Chinese New Year; both sides can go to their own parents' house for the festival but cannot be together. Therefore, the payoff cancels out the loss, and both sides get '0'. Thus, it can be seen that if the wife has already chosen to go to the wife's home, then the husband will definitely choose the 'wife's home'. Thus for the wife, the payoff for choosing the 'wife's home' is '2'.

Taking the above together, if the wife makes a first move, then she should choose the 'wife's home'. Thus, the game outcome is that the payoff for the wife is 2 while the husband's payoff is 1. It can be seen that because the wife makes the first move, her payoff is bigger than the husband's.

On the contrary, if the man makes the first move, then the game is as in Fig. 4.2.

With inverse method, it can be seen that the equilibrium point of this game is where both the wife and husband choose the husband's home. The outcome is that the payoff for the wife is 1, the payoff for the husband is 2.

Similar to the above, where the husband makes the first move, the process with inverse method is:

Firstly since the husband has already chosen the 'husband's home', let us see how the wife will choose. It can be seen in the top part of Fig. 4.2 that if the wife chooses the 'husband's home', the result is husband and wife both go to the husband's home for the Chinese New Year, then the payoff for the husband is '2', the wife '1'. But if the wife chooses the 'wife's home', then the outcome is that the husband goes home for the festival, the wife goes home for the festival, and both sides can go to their own parents' house but cannot be together. Therefore, the payoff cancels out the loss, and

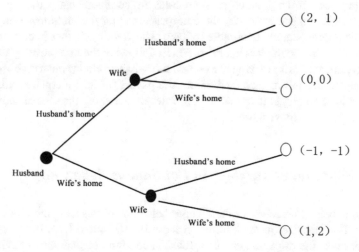

Fig. 4.2 The husband makes the first move in a husband and wife game

both sides get '0'. Thus, it can be seen that if the husband has already chosen to go to the husband's home for the Chinese New Year, then the wife will definitely choose the 'husband's home'. Thus, for the husband, the payoff for choosing the 'husband's home' is '2'.

Since the husband has already chosen the 'wife's home', let us see how the wife will choose. It can be seen in the lower part of Fig. 4.2 that if the both the husband and wife choose the 'wife's home', the result is husband and wife together go to the wife's home, then the payoff for the husband is '1', the wife '2'. However, if the wife chooses the 'husband's home', then the outcome is the husband goes to the wife's home to spend the Chinese New Year, and the wife goes to the husband's home. Neither side can go to their own parents' house, nor can they be together. Therefore both the husband and wife get '−1'. Thus, it can be seen that if the husband has already chosen to go to the wife's home for the Chinese New Year, then the wife will definitely choose the 'wife's home'. Thus, for the husband, the payoff for choosing the 'wife's home' is '1'.

Taking the above together, if the husband makes the first move, then he should choose the 'husband's home'. Thus, the game outcome is that the payoff for the wife is 1, and the husband's payoff is 2. It can be seen that because the husband makes the first move, his payoff is bigger than the wife's.

Note that in this example of the husband and wife game, we have defined the two parties as 'rational persons', i.e. the criterion for both sides in choosing an action is to maximise one's own interests. In such a situation, the 'husband and wife' are aiming for their own payoff and not in looking after the other. If both sides aim to maximise the other's payoff (this is a selfless couple in such a situation), then the game outcome is entirely different.

The husband and wife game can be used to describe a relationship between enterprises and how their products correlate. Often there is a first-move advantage between these enterprises. If the products on both sides are complementary, then the payoff for both will be bigger. However, the enterprise making the first move can choose to produce products more favourable to itself. Faced with whether to choose 'complementary' or 'non complementary' products following the first move by the other enterprise, the enterprise making the second move has no choice but to choose 'complementary products'. In such a situation, the payoff for the enterprise making the first move is often bigger than that for the enterprise making the second move, thus leading to the 'first-move advantage'.

4.1.2   First-Move Advantage in Continuous Behaviour

Continuous behaviour refers to the change between actions as being a continuous situation. For instance, in the bidding process if each quote from the buyer is an action, because the price that can be quoted is continuous, then the behaviour is a kind of continuous behaviour.

In order to contrast it with the advantage of the person making the first move, we will first introduce a classic game model without first-move advantage—the Cournot model. Then the model with first-move advantage will be introduced—the Stackelberg model.

Example 4.2   The Cournot model—simultaneous decision-making (this example is taken from *Game Theory and Information Economics* by Zhang Weiying).

The Cournot model is also called the Cournot duopoly model, put forward by the French economist Antoine Augustin Cournot in 1838. This model describes how the output from two manufacturers, in producing a homogeneous product, reach equilibrium when there is no coordination between them.

Through a high degree of abstraction of reality, an analysis of the problem is made simple and easily understood by using the Cournot model.

The condition of a Cournot model is:

A certain product is manufactured by two enterprises 1 and 2; the production cost of that product is zero.

The function of the product's market price is $P = a - (q_1 + q_2)$. Within this, P is the product's selling price, a is the constant, q_1 and q_2 are the output of enterprises 1 and 2. Both enterprises accurately understand the demand curve in that market. Both enterprises first estimate the other's output amount, before determining their own output that can bring the biggest profit.

More importantly, when the two enterprises decide on their own output, it must be done simultaneously without the situation of one enterprise determining its output before the other.

Let us suppose that:

q_1 is enterprise 1's output, q_2 is enterprise 2's output, π_1 is enterprise 1's profit, π_2 is enterprise 2's profit, then:

$$\pi_1 = q_1 P = q_1[a - (q_1 + q_2)]$$

$$\pi_2 = q_2 P - q_2[a - (q_1 + q_2)]$$

Seeking the optimal output for enterprise 1 and 2:

$$\frac{\partial \pi_1}{\partial q_1} = a - (q_1 + q_2) - q_1 = a - q_2 - 2q_1 = 0$$

$$\frac{\partial \pi_2}{\partial q_2} = a - (q_1 + q_2) - q_2 = a - q_1 - 2q_2 = 0$$

Solving the optimal output for enterprise 1 and 2:

$$q_1^* = \frac{1}{2}(a - q_2)$$

$$q_2^* = \frac{1}{2}(a - q_1)$$

Everything being equal between the two enterprises, solving the above simultaneous equations will arrive at the Nash Equilibrium of the simultaneous decisions on output by the two enterprises:

$$q_1^* = q_2^* = \frac{a}{3}$$

Example 4.3 Stackelberg's leader-follower model—first-move advantage (this example is taken from *Game Theory and Information Economics* by Zhang Weiying).

Heinrich Freiherr von Stackelberg, 1905–1946, was a German economist. In 1934 he published the book *Marktform und Gleichgewicht* in which he put forward the well-known leader-follower model. Later it was called the Stackelberg leadership model, i.e. the Stackelberg leader-follower model.

In the Stackelberg leader-follower model, the game players are enterprises that are leaders and followers. The game is still about choosing the output for the same product. The two sides' actions in order are: The leader first decides on his own output; after the follower has seen the leader's decision, he decides on his own output. The leader knows that the follower will observe this output decision and devise his own output based on having seen the leader's results. Therefore the leader must fully consider this when making decisions.

Let us suppose that enterprise 1 is the leader, its output is q_1; enterprise 2 is the follower, its output is q_2. Let us also suppose that the product's demand price function is still $P = a - (q_1 + q_2)$. In order to compare this with the above Cournot model, here let us still assume that the production cost of the product is 0.

Since the Stackelberg leader-follower model describes the order of game actions by the two parties, this can be solved by using backward induction. For this, let us first find the output decision of enterprise 2 who act last. When the output q_1 of enterprise 1 is known, enterprise 2's payoff function is:

$$\pi_2 = q_2 P = q_2[a - (q_1 + q_2)]$$

Find the derivation and make it 0, we have:

$$\frac{\partial \pi_2}{\partial q_2} = a - (q_1 + q_2) - q_2 = a - q_1 - 2q_2 = 0$$

The solution is: $q_2^* = \frac{1}{2}(a - q_1)$
For enterprise 1, its payoff function is:

$$\pi_1 = q_1 P = q_1[a - (q_1 + q_2)].$$

Substituted by $q_2^* = \frac{1}{2}(a - q_1)$:

$$\pi_1 = q_1 P = q_1[a - \left(q_1 + \frac{1}{2}(a - q_1)\right)] = q_1[a - \left(q_1 + \frac{1}{2}a - \frac{1}{2}q_1\right)]$$

$$= q_1[a - \left(\frac{1}{2}q_1 + \frac{1}{2}a\right)] = q_1(\frac{1}{2}a - \frac{1}{2}q_1) = \frac{1}{2}(aq_1 - q_1^2)$$

Find the derivation and make it 0, we have:

$$\frac{d\pi_1}{dq_1} = \frac{1}{2}(a - 2q_1) = 0$$

The solution is:

$$q_1^* = \frac{a}{2}$$

Substituting it into $q_2^* = \frac{1}{2}(a - q_1)$, we have

$$q_2^* = \frac{1}{2}(a - q_1) = \frac{1}{2}\left(a - \frac{a}{2}\right) = \frac{a}{4}$$

Therefore, enterprise 1's biggest payoff is:

$$\pi_1 = \frac{1}{2}(aq_1 - q_1^2) = \frac{1}{2}(a\frac{a}{2} - \frac{a^2}{4}) = \frac{1}{2}(\frac{a^2}{4}) = \frac{a^2}{8}$$

$$\pi_2 = q_2[a - (q_1 + q_2)] = \frac{a}{4}[a - \left(\frac{a}{2} + \frac{a}{4}\right)] = \frac{a}{4}(\frac{a}{4}) = \frac{a^2}{16}$$

Contrasting the Stackelberg model with the Cournot model, when the two enterprises act at the same time, the Nash Equilibrium is $q_1^* = q_2^* = \frac{a}{3}$, whereas when the two enterprises act in sequence, then the output of enterprise that acts first is $q_1^* = \frac{a}{2}$, and the output of the enterprise that acts later is $q_2^* = \frac{a}{4}$. In comparing these two equilibria results, we discover that in the Stackelberg model, enterprise 1's equilibrium quantity is bigger than that in the Cournot model; enterprise 2's equilibrium quantity is smaller than that in the Cournot model; enterprise 1's profit is bigger than that in the Cournot model; enterprise 2's profit is smaller than that in the Cournot model. Enterprise 1 made the first move and so has a certain advantage—this is the 'first-move advantage'.

In commercial competition, companies in high-tech industries often seek to win by continuous development of new products. In fact, this is the first-move advantage in games. When a new product appears, those who copy it later need a certain period of time before establishing a production scale. Therefore the first mover can monopolise during that period. Also, once a new product has formed brand and customer loyalty, followers will have to pay a high cost to break into the market.

In reality, software development, developing technical standard, setting traffic regulations, computer keyboard layout, and even choosing a country's official language all involve first-move advantages.

4.2 Second-Move Advantage

Do all dynamic games have first-move advantage? The answer is negative. Some games have second-move advantage.

Let us look at a game: There are 3n (n equals 1) cards, and two persons take turns to pick cards in a game. Rules: Each person can pick 1 or 2 cards at each turn. The person who picks the last card wins.

For this game, the winner must be the second mover. The reason is that the second mover can decide whether to pick one or two cards based on the number of cards left behind, always leaving 3m (m is smaller than n) cards for the other. For instance, if there are 7 cards left, one will be picked, leaving 6 for the other; if 5 cards are left, 2 will be picked, leaving 3 for the other. The result is that whether the other picks 1 or 2 cards, the last card will always be his.

With regards to the 'price setting' game between enterprises producing a homogeneous product, the second-move advantage is also true. For so long as the price set by the other enterprise is known, this enterprise can set the product price slightly lower than the other side to attract consumers.

Table 4.1 shows that in the game (this example is taken from *Introduction to Game Theory* edited by Wang Zeke and Li Jie), player 1 has the second-move advantage: If player 2 chooses an action first, he will definitely choose 'left' ('right' is a weak and poor action and should not be chosen). Thus, player 1 can choose 'up', so that both sides reach (4.12) Nash equilibrium, and the payoff is 4.

In contrast, if player 1 makes a move first, he will choose 'up'. The result is that it will make no difference whether player 2 chooses 'left' or 'right'. Now, player 1 will face a bigger risk.

Table 4.1 A game where player 1 has second-move advantage

Player 2

		Left	Centre	Right
Up		**12**	10	**12**
		4	<u>3</u>	2
Player 1		<u>12</u>	10	11
Down		<u>3</u>	2	1

Table 4.2 An example of both first-move advantage and second-move advantage

| | | Player 2 | |
		Left	Right
Player 1	Up	0 10	4 5
	Down	100 10	0 5

In economics theories there is a 'theory on order' in economics or the 'entry order effect'. That theory mainly investigates whether the first mover or the second mover has the advantage in competitive markets. Regardless of the various views, on the whole, that theory leans towards 'second-move advantage'. For instance, Lieberman and Montgomery (1990) believe that there is second-move advantage in at least three aspects:

(1) The 'free ride' effect: The second mover can probably save more than the first mover in investment for product design and research, fostering customer loyalty, staff training, government inspection, infrastructure etc.
(2) The first mover is at a higher risk for errors regarding technology or marketing strategy because there is almost no experience to learn from.
(3) The first mover has to pay dearly for technology and experience which can easily be copied.

4.3 Both First-Move Advantage and Second-Move Advantage May Be Present

In sequential games, some are of neither first-move advantage nor second-move advantage, some are of first-move advantage, some are of second-move advantage, while for others both first-move advantage and second-move advantage are present.

Please see the game in Table 4.2 (this example is taken from *Introduction to Game Theory* edited by Wang Zeke and Li Jie).

In this game, player 1 has first-move advantage. That is, he chooses 'down' first and is bound to get 10. If player 2 is allowed to make a first move by choosing 'left', the risk for him is very high because in such a situation, if by chance player 1 chooses 'up', then player 2 can only achieve 0. Therefore, to be on the safe side, player 2 will generally choose 'right' first so that player 1 can only achieve 5.

The characteristic of this game is that while player 1 has the first-move advantage, player 2 also has the second-move advantage. He can let player 1 make a first move; player 1 will definitely choose 'down' so that he can choose 'left' and ultimately achieve the big payoff of 100.

Chapter 5
Credible Commitment and Credible Threat in Games

5.1 Credibility of Commitments and Threats

5.1.1 Commitment and Threat

Commitment and threat are important subject matters in game theory.

A commitment is a pledge advantageous to the other party, whereas a threat is a pledge disadvantageous to the other party.

T. C. Schelling, the Nobel Prize winner in economics, defined commitment and threat as: A announces that B's behaviour will lead to a response from A. If this response is a reward, then the announcement is a commitment; if this response is a penalty, then the announcement is a threat.

From the perspective of the recipient of the commitment or threat, judging from the effect on the payoff for the party who issues the commitment and threat in carrying out the pledges, some commitments and threats are credible, while others are not credible.

If in carrying out the pledges under pre-agreed conditions is more advantageous to the party who issues the commitment or threat, then these pledges are credible. By contrast, if in carrying out the commitment or threat it is not of advantage or is even harmful to the party that issues the pledges, then these pledges are not credible.

In a game, the objective of the commitment and threat affect the other party's choice of behaviour.

5.1.2 Credibility of Commitments and Threats

In reality, some commitments and threats are credible, some are not.

Here, the principle for judging the credibility of the commitment and threat is that, given the conditions that can prompt the commitment and threat, if it is more

© China Economic Publishing House and Springer Nature Singapore Pte Ltd. 2018
S. Sun and N. Sun, *Management Game Theory*,
https://doi.org/10.1007/978-981-13-1062-1_5

advantageous to carry out the pledges by the person who issued the commitment or threat, then the commitment or threat is credible. By contrast, if it is disadvantageous to the person who issued the commitment or threat to carry out the pledges, then the commitment or threat is not credible. This is because we believe that a rational person, when faced with choices of behaviour, will only choose the one that is most beneficial to himself. It must be emphasised here that the research in the credibility of commitment and threat is based on the assumption of a rational person. It is not possible to judge the credibility of commitment and threat of an irrational person.

Example 5.1 Threats of the trade union

When a trade union demands a pay rise for the workers of an enterprise, the threat often used is: "If we do not get a pay rise we will go on strike." For the enterprise, this is a non-credible threat. Although the enterprise will suffer a loss in a strike, so will the workers. The condition that the enterprise 'does not give workers a pay rise' will trigger the workers to choose the 'no strike' behaviour which is more beneficial to them. It should be pointed out that 'non-credible' here does not mean it will 'definitely not happen'. For instance, if the trade union believes that the enterprise will finally find it too much and will have to give the workers a pay rise in the end, then the trade union will choose to 'strike'.

5.2   How to Increase the Credibility of Commitments and Threats

During a game the way to make one's commitment or threat credible is to change the environment that will trigger the condition so that carrying out the commitment or threat becomes one's advantageous strategy. In such a situation, when the condition is triggered, the other party, in assuming that the opponent in this game is rational, will believe that the pledge will definitely be carried out.

Schelling, who received the Nobel Prize in economics for his research in credible commitments, cited an example: A tailor wants to borrow some money from someone. It is not a large sum of money and if he does not pay it back, he will not be penalised legally. Here, if he simply says 'I will definitely pay it back. Please believe me', then this is a non-credible commitment because in such a situation it is more advantageous to him not to pay back the money (not taking into account the loss in reputation and the adverse effect it will have on his business). By contrast, if he puts his sewing machine up as collateral for the creditor (assuming the market value of his sewing machine is greater than the amount he wants to borrow), then his promise to pay back the money is a credible commitment. It is credible because if he defaults, the creditor keeps the collateral and he will suffer greater loss.

In pledging collateral for a loan, paying a deposit for ordering goods, etc., people are in fact turning non-credible commitments into credible commitments. The ancient Chinese idiom on the story to 'break the cauldrons and sink the boat' (meaning to burn one's bridges) is in fact to make carrying out the threat inevitable by taking into account all the choices one has once the conditional option is changed.

Studying the credibility of a commitment and how to increase its credibility has significant meanings in sectors such as commercial contracts.

5.3 Both Parties in a Game Have no Credible Threat—The Chicken Game

In game theory publications there is a well-known chicken game. 'Chicken' in American slang means a 'coward', so the chicken game means a 'coward's game'.

The original chicken game comes from an American film in the 1950s. There is this scene in the film: Two drivers were testing each other's nerve and the rules for the contest were agreed beforehand: The two will drive simultaneously towards each other on a collision course. The one who swerves at the last minute to avoid a crash will be the loser. Of course, if neither swerves, then the cars collide and there will be casualties.

In this game, if either side proposes that he 'will not yield', it is a non-credible threat. This is because if the threatened side does not believe it and is determined to go ahead, the side issuing the threat has two choices: One is to make good and carry out the threat, i.e. to crash towards the other; the other is to swerve with the result to save his own life. Clearly, for a completely rational person, if the other side crashes towards you recklessly, it is more advantageous to swerve rather than to crash ahead.

In the chicken game, if one side is reckless and irrational whereas the other side is completely rational, then because the 'irrational person' does not think about the consequences, the rational person cannot judge if the other side will choose 'avoidance'. However, knowing that the other side is 'irrational', he will more than likely judge that the other side will 'recklessly crash forward', therefore he can only choose to 'avoid'. In such a situation, the irrational person is often the winner in the game.

5.4 The Market Entry Game—One Side Has no Credible Threat

The structure of the market entry game is as shown in Fig. 5.1. That game has two players—enterprise 1 is an oligopoly that has hold of the market of a certain product, and enterprise 2 plans to challenge enterprise 1 by producing that product and so enters the market.

The process of the game for both sides is:

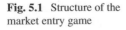

Fig. 5.1 Structure of the market entry game

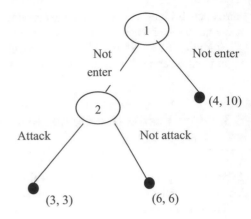

Under the situation that enterprise 2 has hold of the whole market, it discovers that enterprise 1 wants to enter the market. Because of this, enterprise 2 sends a threat to enterprise 1: If you enter the market, I will hit you by reducing the product price.

Under such a threat from enterprise 2, enterprise 1 has to decide whether to 'enter the market' or 'not enter the market'. If it does not enter the market, the game is over for both sides; enterprise 2 still has hold of the whole market and its payoff is 10. Since enterprise 1 does not enter the market but continues its business in its own area, its payoff is 4.

If enterprise 1 enters the market, then enterprise 2 will decide whether to 'hit enterprise 1 with a price reduction' or 'no price reduction so as not to fight enterprise 1'.

If enterprise 2 does not fight with enterprise 1, then enterprise 1's payoff is 6, and because of enterprise 1's entry, the market is under attack so that enterprise 2's payoff is reduced from the original 10 to 6.

If enterprise 2 attacks enterprise 1, then enterprise 1 will suffer a loss with a payoff of 3, thus it might as well not enter the market. Having to fight enterprise 1 with a price reduction, enterprise 2's own payoff is affected, reduced from the original 10 to 3.

Clearly, enterprise 2's threat to enterprise 1 with 'if you enter the market I will hit you by reducing the product price' **is a non-credible threat**. Under the situation that enterprise 1 decides to 'enter', the best action for enterprise 2 is 'not attack' rather than 'attack'. Therefore, the equilibrium outcome of this game is (enter, not attack).

5.5 Burning One's Bridges: One Side with Credible Threat

From the perspective of game theory, the Chinese idiom of 'break the cauldrons and sink the boat' in fact illustrates the situation where 'one party has a credible threat'. During the Qin and Han dynasties in China, the Qin army surrounded the state of Zhao. Xiang Yu, the leader of the farmer's insurrection, led his forces to assist the state of Zhao. At the time, the Qin army was very strong. Xiang Yu's men were not willing to fight out of fear of defeat. By now, Xiang Yu must find a way to make his men fight hard, which got him into a game situation with his men.

To make his men put aside the thought of life or death, Xiang Yu made a threat to his men that 'not fighting hard will mean death':

Xiang Yu led the whole army across the Zhang River, all the boats were sunk, pots and pans were all smashed, the entire military camp was burnt, and only took three days' supply of food, so as to show the soldiers that they must be determined to fight till they die; there was to be no thought of withdrawing. The army got to the front line and surrounded Wang Li. The army then faced the Qin army and fought many battles, blocking the road built by the Qin army and wiped them out. Su Jiao was killed and Wang Li was captured.

This game in essence is that after the soldiers crossed the Zhang River, Xiang Yu destroyed all the boats. In such a situation, if the soldiers did not fight hard, Xiang Yu did not have the choice of 'letting the soldiers withdraw'. Thus, Xiang Yu's threat that 'if they do not fight hard they will die' is a credible threat.

Chapter 6
Coalitional Games

6.1 Coalitions

6.1.1 Grand Coalition

Assume N is a set formed by n persons, that is $N = \{1, 2, \ldots, n\}$.

In cooperative games, the above n persons that form the coalition is called a grand coalition.

People form coalitions because they want to achieve payoffs greater than those they can get on their own. Only when a coalition offers payoff that is superadditive can it make each individual's payoff exceed that of individuals going it alone.

For this, it is necessary to define additivity, superadditivity, and subadditivity.

Additivity

For any two minor coalitions (including situations with only one player) S, T, if $S \cap T = \varnothing$, then $v(S \cup T) = v(S) + v(T)$. Then, that coalition is an additive coalition. For an additive coalition the payoff is the same whether or not an individual joins the coalition (that is, if each individual's allocation does not encroach on other's contribution).

Superadditivity

For any two minor coalitions S, T, if $S \cap T = \varnothing$, then $v(S \cup T) > v(S) + v(T)$. Then that coalition is a superadditive coalition. Clearly, a superadditive coalition benefits from size and scale. A coalition's superadditivity is the motivation for individuals to join the coalition.

Subadditivity

For any two minor coalitions S, T, if $S \cap T = \varnothing$, then $v(S \cup T) < v(S) + v(T)$. Then that coalition is a subadditive coalition. Clearly, in a subadditive coalition, the

© China Economic Publishing House and Springer Nature Singapore Pte Ltd. 2018
S. Sun and N. Sun, *Management Game Theory*,
https://doi.org/10.1007/978-981-13-1062-1_6

payoff for individuals is reduced after joining the coalition. For instance, in groups featuring serious internal conflicts and mutual constraints, the payoff for individuals is not as good as those individuals who go it alone.

6.1.2 Number of Sub-coalitions

For $N = \{1, 2, \ldots, n\}$, any coalition with persons fewer than or equal to n (including coalitions with no one), it is called a sub-coalition.

Now let us consider the possible number of sub-coalitions.

Firstly, an individual 'going it alone' is the smallest sub-coalition. For $N = \{1, 2, \ldots, n\}$, there are $\begin{bmatrix} n \\ 1 \end{bmatrix} = n$ sub-coalitions of this kind.

Next, a sub-coalition can be formed of two persons; there are $\begin{bmatrix} n \\ 2 \end{bmatrix}$ 'two-person sub-coalitions'.

Likewise, the number of 'three-person coalition' is $\begin{bmatrix} n \\ 3 \end{bmatrix}$. The number of '$k$-person coalition' is $\begin{bmatrix} n \\ k \end{bmatrix}$ $(1 \le k \le n)$.

6.1.3 The Number of All Coalitions

Furthermore, from the perspective of taking into account all possibilities, the unusual sub-coalition of an 'empty coalition' $\varnothing$ should be considered. $\varnothing$ containing no one, there is only $\begin{bmatrix} n \\ 0 \end{bmatrix} = 1$ empty coalition.

Since a sub-coalition is defined as 'any coalition with persons fewer or equal to n', grand coalitions (including those of n players), empty coalitions (these two coalitions can be treated as special sub-coalitions) and any other possible sub-coalitions can come to a total number of

$$\begin{bmatrix} n \\ 0 \end{bmatrix} + \begin{bmatrix} n \\ 1 \end{bmatrix} + \begin{bmatrix} n \\ 2 \end{bmatrix} + \cdots + \begin{bmatrix} n \\ n \end{bmatrix}$$

For the summation of the above formula, let us look at the binomial formula:

$$(x + y)^n = \begin{bmatrix} n \\ 0 \end{bmatrix} x^0 y^n + \begin{bmatrix} n \\ 1 \end{bmatrix} xy^{n-1} + \cdots + \begin{bmatrix} n \\ k \end{bmatrix} x^k y^{n-k} + \cdots \begin{bmatrix} n \\ n \end{bmatrix} x^n y^0$$

Making $x = y = 1$, we have:

$$\begin{bmatrix} n \\ 0 \end{bmatrix} + \begin{bmatrix} n \\ 1 \end{bmatrix} + \begin{bmatrix} n \\ 2 \end{bmatrix} + \cdots + \begin{bmatrix} n \\ n \end{bmatrix} = 2^n$$

Thus, for $N = \{1, 2, \ldots, n\}$, the number of all sub-coalitions that can be formed (including the grand coalition itself, empty coalitions and various 'real' sub-coalitions) is 2^n.

For this reason, publications of cooperative games often mark all possible sub-coalition groupings as 2^N (note that this is only a set symbol, used to indicate that it contains a set of 2^n elements). Thus, any sub-coalition can be represented as $S \in 2^N$. And $|S|$ represents the number of members in coalition S.

6.1.4 Characteristic Vector

A characteristic vector is used to accurately represent and differentiate 2^n the number of possible sub-coalitions.

Suppose the n dimensional characteristic vector that indicates S is marked as e^s, then its i element is.

$$\left(e^S\right)_i = \begin{cases} 1 \text{ When } i \in S \\ 0 \text{ When } i \in N\backslash S \end{cases} \tag{6.1}$$

In Formula (6.1), $N\backslash S$ shows from N the subset S is removed, and we get the complement.

Clearly, the characteristic vector of a grand coalition is $e^N = (1, 1, \ldots, 1)$.

6.1.5 Characteristic Function

For $S \in 2^N$, use the set-valued function $v(S)$ to represent coalition S's payoff for all its members taken together. To differentiate it from a general algebraic function, the independent variable of the set-valued function is 'set' instead of a numerical variable in general functions; therefore the variable $v(S)$ is an actual number.

The set-valued function $v(S)$ that reflects the coalition's payoff is called the characteristic function. It represents the transferable utility or payoff that the coalition can obtain. It should be emphasised that $v(S)$ is coalition S's 'bottom line' for achievable payoff, i.e. any coalition in $N\backslash S$ cannot cause the payoff for S to be less than $v(S)$.

Note that S is a set of players, so for an enumeration of all the elements in S, the regular set symbol is $S = \{1, 2, \ldots, k\}$. Thus, the enumeration of the characteristic function should be shown as $v(\{1, 2, \ldots, k\})$; but this appears very cumbersome, so in this book $v(1, 2, \ldots, k)$ is used to represent $v(\{1, 2, \ldots, k\})$. For instance $v(i)$ is used to represent $v(\{i\})$.

Example 6.1 The characteristic function in the Airport Game (Littlechild and Owen 1973) (this example is taken from *Cooperative Game Theory* by Dong Baomin, Wang Yuntong and Guo Guixia, and *An Introduction to Cooperative Games* by Shi Xiquan).

n Airline companies $N = \{1, 2, \ldots, n\}$ own different sizes of aircrafts, and each of these airlines only has aircrafts of a certain size. Large aircraft needs a longer runway than small aircraft. Runways that can accommodate large aircraft to take off and land will definitely accommodate small aircraft.

Suppose for aircraft size i, the construction cost of the runway is $c_i (i = 1, 2, \ldots, n)$. We can also suppose the size of aircraft is in this order $1 \leq 2 \leq \cdots \leq n$. Thus, the construction cost of runways for each size of aircraft alone is $c_1 \leq c_2 \leq \cdots \leq c_n$. Thus, for the coalition $S \in 2^N$ formed by some airlines, the cost outlay is only the runway construction cost for the biggest aircraft. Therefore, the cost borne by all airlines in coalition S is:

$$v(S) = \max\{c_p, c_q, \ldots, c_w\} \quad (p, q, \ldots, w) \in S \tag{6.2}$$

Readers should note that in this example, $v(S)$ is the cost, not the payoff.

Example 6.2 The characteristic function of the Glove Game (this example is taken from *An Introduction to Cooperative Games* by Shi Xiquan).

In a crowd $N = \{1, 2, \ldots, n\}$ there exists two partitioned subsets L and R ('partitioning' results in subsets L and R both fulfilling the conditions $L \cap R = \varnothing, \ L \cup R = N$. In L each person only owns one left hand glove; in R each person only owns one right hand glove. The value of a single glove is 0; the value of a pair of matching left and right gloves is 100 yuan.

In such a situation, for any sub-coalition S, the more gloves that can be paired up, the bigger the payoff for the coalition.

Thus, for any one $S \in 2^N$, S can include members of subset L and members of subset R. Therefore, the number of gloves that can be paired up is the smallest number in subset $|S \cap L|$ (number of people holding a left hand glove in S) and subset $|S \cap R|$ (number of people holding a right hand glove in S).

Therefore, the characteristic function of the Glove Game

$$v(S) = 100 \times \min\{|S \cap L|, |S \cap R|\}, \forall S \in 2^n \tag{6.3}$$

For instance, there are 20 people in S, of which 12 people have a left-hand glove, eight have a right-hand glove, then $v(S) = 100 \times \min\{12, 8\} = 100 \times 8 = 800$ yuan.

6.1.6 *Definition of Cooperative Games*

Cooperative games can be defined by the set of players N and the characteristic function v.

Definition 6.1 A cooperative game is a 2-tuple $\langle N, v \rangle$, having N as the set of players, v as the characteristic function, and $v : 2^N \rightarrow R$, i.e. a mapping from set 2^N to the real number set R.

Normally, G^N is used to indicate the cooperative game with N as the set of players. For determining the set of players only, there can be many kinds of characteristic functions. Hence $v \in G^N$ is often used to represent a certain characteristic function with N as the set of players.

6.2 Imputation and the Core

For cooperative games of transferable utility, the 'solution of the game' is to find a scheme acceptable to all for allocating collective payoff between the players (or sharing cost). Only if such an allocation scheme can be found and executed can the 'coalition' be maintained and not disintegrate.

Clearly, for coalitional games, the 'solution of the game' is a scheme for allocating collective payoff (or a scheme for sharing collective cost). It must be satisfactory to each of the player as well as each of the sub-coalition. This is because, if any of the sub-coalition (including the smallest sub-coalition—an individual) is dissatisfied with the allocation, they will leave the coalition and it will disintegrate.

Example 6.3 (Imma Curiel 1997) There are five people (marked as 1, 2, 3, 4, 5 respectively) in different situations: they may have funds, technical know-how, factory premises, personal connections. These five people decide to partner up to set up a business. The business feasibility report states that when the enterprise is set up, the annual profit will be 1 million US dollars. To enable the five people to cooperate, it is necessary to determine a reasonable scheme for allocating the annual profit (in million US dollars) (Table 6.1).

Table 6.1 Possible cooperation between the five and the corresponding profits

S	V(S)	S	V(S)	S	V(S)	S	V(S)
{1}	0	{1, 5}	20	{1, 2, 4}	35	{3, 4, 5}	70
{2}	0	{2, 3}	15	{1, 2, 5}	40	{1, 2, 3, 4}	60
{3}	0	{2, 4}	25	{1, 3, 4}	40	{1, 2, 3, 5}	65
{4}	5	{2, 5}	30	{1, 3, 5}	45	{1, 2, 4, 5}	75
{5}	10	{3, 4}	30	{1, 4, 5}	55	{1, 3, 4, 5}	80
{1, 2}	0	{3, 5}	35	{2, 3, 4}	50	{2, 3, 4, 5}	90
{1, 3}	5	{4, 5}	45	{2, 3, 5}	55	{1, 2, 3, 4, 5}	100
{1, 4}	15	{1, 2, 3}	25	{2, 4, 5}	65	∅	0

This example is taken from *Cooperative Game Theory* by Dong Baomin, Wang Yuntong, and Guo Guixia

Firstly, someone propose that the profit should be equally distributed (i.e. each person gets 200,000 US dollars).

However, players 4 and 5 find that if just the two of them cooperate, their annual payoff can reach 450,000 US dollars, great than the 400,000 US dollars payoff they can get in the five-person cooperation. Therefore, in view of the scheme for equal distribution, 4 and 5 propose that they do not join the five-person cooperation but the two of them cooperate instead.

The remaining people 1, 2 and 3 find that if only the three of them cooperate, the annual payoff is only 250,000 US dollars. Therefore, 1, 2 and 3 decide to find ways to keep 4 and 5 in. Thus, they decide to give 460,000 US dollars to 4 and 5 (more than the 450,000 US dollars created by the cooperation between 4 and 5), and then equally share out the remaining 540,000 US dollars between 1, 2 and 3.

But then a new problem arises: Players 3, 4 and 5 find that if the three of them leave the grand coalition to cooperate by themselves, the annual payoff is 700,000 US dollars, more than the 640,000 US dollars when 4 and 5 cooperate (460,000 US dollars+180,000 US dollars).

If this happens, only 1 and 2 are left to cooperate together; due to the lack of the funds and know-how, the payoff for their cooperation is 0. Thus, 1 and 2 can only agree to apportion 710,000 US dollars between 3, 4, and 5, the remaining 290,000 US dollars is equally shared between 1 and 2.

Obviously, for transferable utility in cooperative games, to seek a 'solution for the game' is to find an allocation scheme satisfactory to all the sub-coalitions. This allocation scheme can be expressed as the payoff vector $x = (x_i)_{i \in N} \in R^n$ constituted by the payoff of all the players, of which x_i is cooperation payoff allocated to player i.

Below, we will explore the conditions that are necessary in an allocation scheme that can satisfy all the sub-coalitions.

6.2.1   Collective Rationality—The Feasibility of the Allocation Scheme and the Player's Degree of Satisfaction

Firstly, from the perspective of the feasibility of the allocation scheme, only if $\sum_{i \in N} x_i \leq v(N)$'s payoff vector is satisfied can it be feasible. In other words, after allocating the payoff to the players, the coalition should not be in deficit.

On the other hand, from the perspective of the players, clearly the greater the payoff allocated, the greater the satisfaction. Considering these two conditions together, the allocation scheme should satisfy the equation as follows:

$$\sum_{i \in N} x_i \leq v(N)$$

This condition that must be satisfied in an allocation scheme is called the collective rationality.

6.2.2 Individual Rationality—Condition for an Individual to Join a Coalition

For certain individuals, they can choose to join to a coalition and also to go it alone. Whether to join a coalition or to go it alone depends on the size of the payoff for these two choices. Suppose for the individual $i \in N$, his payoff for working alone is $v(i)$, and the payoff for joining a coalition (with the coalition adopting a certain allocation scheme) is x_i. Clearly, when $x_i < v(i)$, i will not join the coalition but work alone. Only when it is $x_i \geq v(i)$, an individual will join a coalition. In other words, for the set of players N, when the allocation scheme ensures that all the people in N can have their payoff as $x_i \geq v(i)$ in the coalition, then this N-person coalition can be sustained. This condition that must be satisfied in an allocation scheme is called the 'individual rationality'.

6.2.3 The Imputation Set

Definition 6.2 If the allocation scheme (payoff vector) $x \in R^n$ **can simultaneously satisfy the collective rationality and individual rationality,** i.e. $\sum_{i \in N} x_i = v(N)$ and also $x_i \geq v(i)$ is true for all $i \in N$, then x is an imputation of the game $v \in G^N$.

Definition 6.3 All the imputations of the game $v \in G^N$ form **the imputation set, which** is marked as $I(v)$.

Obviously, if the imputation set $I(v)$ is not empty, then it is possible to choose an allocation scheme from $I(v)$ to sustain the coalition N.

But the question is under what conditions is $I(v)$ not empty? In this regard, it is evident from published works that $v(N) < \sum_{i \in N} v(i)$ is a necessary and sufficient condition for $I(v)$ as an empty set.

In fact, $v(N) < \sum_{i \in N} v(i)$ means that the coalition's payoff is smaller than the arithmetic sum of all the players' 'going-it-alone payoff'. This means that this so-called 'coalition' is not superadditive: the efficiency is not as high as each individual working alone. This is the case with the 'inefficient coalitions' featuring lots of internal conflicts and mutual constraints.

By contrast, if $v(N) < \sum_{i \in N} v(i)$, it clearly says that this kind of coalition is 'superadditive', i.e. the coalition's payoff exceeds the sum of the payoff of the various individuals working alone. The kind of coalition is able to achieve economies of scale.

Of course, in reality there are coalitions that satisfy the condition $v(N) < \sum_{i \in N} v(i)$. In this kind of coalition, the efficiency generated exactly equals the sum of the payoff of the various players working alone. This kind of coalition is called an 'ordinary coalition', because it makes no difference whether or not an individual joins the coalition, so it does not matter.

With regards to an ordinary coalition, it can be verified that $x = (v(1), v(2), \ldots, v(n))$ is the only imputation.

6.2.4 Small Coalition Rationality

Marking the imputation set as $I(v)$, which is a combination of the allocation scheme that simultaneously satisfies individual and collective rationality.

Now, the new question is, can all the schemes in the imputation set sustain the coalition and prevent it from disintegrating?

Think about the requirement that an imputation has to satisfy individual rationality, which in effect is the basic condition that prevents individuals from choosing to go it alone.

However, if k players $(1 < k < n)$ in N form a small coalition S, and the payoff for this coalition S despite the grand coalition is $v(S)$, which is greater than the sum of the payoff they get from the grand coalition's imputation x, i.e. $v(S) > \sum_{i \in S} x_i$, then this sub-coalition will leave the grand coalition. Thus, the grand coalition N will dissolve because these k players have left.

By contrast, if leaving the grand coalition the sub-coalition S's payoff $v(S)$ is less than the sum of the payoff from the grand coalition's imputation x, i.e. $v(S) > \sum_{i \in S} x_i$, Then in such a situation, the individuals in the small coalition S will not leave the grand coalition.

Definition 6.4 In the set of players N, for any partition $(S_1, \ldots, S_k)$ of N, for each subset S_j $j \in 1, \ldots, k$ after the partition, if $v(S_j) \leq \sum_{i \in S_j} x_i$ exists, then this kind of allocation scheme is small coalition rational.

Since a player by himself is also a subset S_j, small coalition rationality actually also includes individual rationality. In other words, if a certain allocation scheme satisfies the small coalition rationality, then it must also satisfy the individual rationality.

6.2.5 The Core

According to the analysis above, it can be seen that not all the allocation schemes in the imputation set $I(v)$ can definitely guarantee the stability of the coalition. In the imputation set $I(v)$, only the allocation schemes that are small coalition rational can make all the players feel willing to join the grand coalition. This kind of schemes form a subset of $I(v)$, which is called the core, expressed as $C(v)$.

Definition 6.5 The core $C(v)$ in cooperative games is:

$$C(v) = \left\{ x \middle| x \in I(v), \sum_{i \in S} x_i \geq v(S), \forall S \in 2^N \backslash \{\varnothing\} \right\} \tag{6.4}$$

The definition of Formula (6.4) is based on the coalition's allocation of payoff. If cost sharing is carried out within the coalition, i.e. seeking a reasonable cost sharing scheme, then the definition of the core is (6.5):

$$C(v) = \left\{ x \middle| x \in I(v), \sum\nolimits_{i \in S} x_i \leq v(S), \forall S \in 2^N \backslash \{\varnothing\} \right\} \qquad (6.5)$$

In Formula (6.5), $v(S)$ is the cost borne by the sub-coalition alone when it leaves the grand coalition.

According to the definitions of imputation set and core, the difference and relationship between the imputation set and the core can be clearly seen: An imputation set $I(v)$ is a set of allocation schemes that is both individually rational and collectively rational, whereas for a core $C(v)$ the requirements are more strict: It is a set that is individually rational, collectively rational, and small coalition rational. Thus $C(v) \subseteq I(v)$.

6.2.6 The Problem that Exists When the Core Is the Solution to the Coalitional Game

From the perspective of maintaining a coalition's stability, the allocation scheme of the core is without doubt a good choice. However, in practice, the core has shortcomings. There are three aspects to these shortcomings: The allocation scheme is not unique (sometimes there are infinite allocation schemes in a core), the allocation scheme is extreme (in some situations there is serious bias towards the dominant side), and an absence of conditions that meet the allocation scheme (sometimes the core is an empty set).

6.2.6.1 The Allocation Scheme Is not Unique

In many situations the core has many allocation schemes, even infinite. In such a situation, actually choosing an allocation scheme for solving the game becomes difficult.

Example 6.4 An insurers coalition (this example is taken from *An Introduction to Cooperative Games* by Shi Xiquan).

Suppose there are three insurance companies. Insurer i has n_i insured customers ($i = 1, 2, 3$). Each company has a certain probability for claims by their customers. Therefore, insurance companies must have a certain amount of fund in reserve for claims. Let us assume that the premiums paid by customers are all used by insurance companies as reserve for claims. If a certain company has too many customers making claims which exceed the amount in reserve for claims, that insurance company will go bankrupt.

Under certain bankruptcy probability, the greater the number of customers, the smaller the average amount in reserve for claims per customer is needed. This is because according to the probability principle, the greater the number of people, the smaller the probability for claims at the same time.

When $n_1 = 100$, if a customer suffers a loss, then the amount of his loss is 1 (assume it's fully comprehensive cover where the insurer compensates the customer for the full amount of loss, i.e. the amount of loss equals the amount of claim), and the probability of loss is $q_1 = 0.1$. According to the country's regulations, an insurer's amount for claims reserve should be set at a level making sure that the company's bankruptcy probability is less than 0.001 (one thousandth).

Suppose the incident of the customer's loss is independent. Since n_1 is quite big, the normal approximation of the binomial distribution for a large sample can be applied. The amount of reserve for claims company 1 needs is

$$p_1 = n_1 q_1 + 3\sqrt{n_1 q_1 (1 - q_1)} = 10 + 9 = 19$$

Thus, the amount of reserve for claims the company needs is 19. There are currently 100 customers, and $100 \times 0.19 = 19$, therefore each customer has to pay 0.19.

For insurance company 2, $n_2 = 100$. If a customer suffers a loss, the amount of his loss is 1, and the probability for loss is $q_2 = 0.2$. Likewise, an insurer's amount for claims reserve should be set a level making sure that the company's bankruptcy probability is less than 0.001 (one thousandth). The amount of reserve for claims company 2 needs is

$$p_2 = n_2 q_2 + 3\sqrt{n_2 q_2 (1 - q_2)} = 20 + 12 = 32$$

Thus, the amount of reserve for claims the company needs is 32. There are currently 100 customers, and $100 \times 0.32 = 32$, therefore each customer has to pay 0.32.

What is the situation if insurance companies 1 and 2 merge as insurance company 12?

Let us assume that insurance company 12's bankruptcy probability is still less than 0.001, the amount of reserve needed for claims is:

$$p_{12} = n_1 q_1 + n_2 q_2 + 3\sqrt{n_1 q_1 (1 - q_1) + n_2 q_2 (1 - q_2)} = 10 + 20 + 15 = 45$$

Since $p_{12} = 45 < p_1 + p_2 = 51$, the result of merging the two companies is that the amount of reserve for claims is reduced by 6. If the characteristic function is considered (that characteristic function represents cost), it is:

$$v(1) = 19, v(2) = 32, v(1, 2) = 45$$

Suppose insurance company 3 has $n_3 = 120$ customers, the probability for customers making claims is $q_3 = 0.3$, and the amount for a claim is 1. The bankruptcy probability for company 3 is less than 0.001, so the amount of reserve needed for claims is

$$p_3 = n_3 q_3 + 3\sqrt{n_3 q_3 (1 - q_3)} = 36 + 15 = 51$$

Now, the three insurance companies merge to become a large insurance company 123. Under the condition that the bankruptcy probability for company 123 is less than 0.001, the amount of reserve needed for claims is

$$P_{123} = n_1 q_1 + n_2 q_2 + n_3 q_3 + 3\sqrt{n_1 q_1 (1 - q_1) + n_2 q_2 (1 - q_2) + n_3 q_3 (1 - q_3)}$$
$$= 10 + 20 + 36 + 21 = 87$$

$P_{123} = 87 < p_1 + p_2 + p_3 = 102$. It can be seen that the more the customers, the less the amount of reserve for claims per unit of customer.

Suppose the vector of shared expenses for the three insurance companies is (x_1, x_2, x_3). Let us study the solution to this cooperative game below.

First, any allocation scheme in the set of imputations $I(v)$ for that cooperative game must satisfy the conditions below: $x_1 + x_2 + x_3 = 87, x_1 \leq 19, x_2 \leq 32, x_3 \leq 51$.

However, not every scheme in $I(v)$ can keep company 123 from disintegrating. For instance, in $I(v)$ the scheme $(17, 30, 40)$ clearly belongs to $I(v)$. But company 1 and company 2 find that if they form a coalition as company 12, then the fund of reserve for claims is only 45, less than that scheme's $17 + 30 = 47$. Therefore, these two companies can definitely form their own company 12; then company 12's reserve for claims is 45, which can be allocated using the scheme $(16, 29)$, clearly better than the $(17, 30)$ they get in the grand coalition.

In such a situation, company 3 must yield if it wants companies 1 and 2 to cooperate, so that the amount of reserve for claims borne by company 1 and company 2 is less than 45. Thus, the amount of reserve for claims borne by company 3 must be greater than 42 ($x_3 \geq 87 - 45 = 42$).

Likewise, we must also think about the payoffs if company 1 and company 3 form a coalition, or company 2 and company 3 form a coalition. These are:

$$P_{13} = n_1 q_1 + n_3 q_3 + 3\sqrt{n_1 q_1 (1 - q_1) + n_3 q_3 (1 - q_3)} = 10 + 36 + 17.54 = 63.54$$

$$P_{23} = n_2 q_2 + n_3 q_3 + 3\sqrt{n_2 q_2 (1 - q_2) + n_3 q_3 (1 - q_3)} = 20 + 36 + 19.3 = 75.3$$

Thus, to make the allocation scheme (x_1, x_2, x_3) small coalition rational, so that even if any of the companies form small coalitions, their payoff cannot be greater than what they can achieve in the grand coalition, then the allocation scheme in the game's core $C(v)$ must satisfy the conditions below:

$$x_1 + x_2 + x_3 = 87$$

$$x_1 \leq 19, \ x_2 \leq 32, \ x_3 \leq 51$$

$$x_1 + x_2 \le 45, \ x_1 + x_3 \le 63.54, \ x_2 + x_3 \le 75.3$$

Reorganised, we have:

$$x_1 + x_2 + x_3 = 87$$

$$11.7 \le x_1 \le 19, \ 23.46 \le x_2 \le 32, \ 42 \le x_3 \le 51$$

It is not difficult for the readers to find out that the allocation scheme in that game is infinite. As the allocation scheme in the core is not unique, it leaves room for the various parties in their scramble for their interests, but also causes potential problems for the coalition's disintegration.

6.2.6.2 Extreme Allocation Scheme

An extreme allocation scheme refers to the allocation scheme in the core that is sometimes seriously biased towards the dominant party, making it absurd and in fact not feasible.

Example 6.5 The core of Glove Game (this example is taken from *An Introduction to Cooperative Games* by Shi Xiquan).

Suppose $N = \{1, 2, 3\}$, in which 1 owns a left-hand glove, 2 and 3 each owns a right-hand glove. The value of each pair of matched glove is 100 yuan. The value of any single glove is 0. Among the three players, 1 has an important position because without 1's cooperation, no coalition will have any payoff.

What is surprising is that $(100, 0, 0)$ is the only allocation scheme of the 'core' in this game, i.e. 1 has all the payoff, the payoff for 2 and 3 is 0.

This is because if 2 or 3's payoff is greater than 0, then the allocation scheme is unstable.

For instance, let us suppose that the allocation scheme is $(100 - \Delta, \Delta, 0)$, of which $\Delta > 0$. Obviously, player 3 will not be happy with this allocation scheme because his payoff is 0. Now, 3 can propose a new allocation scheme $(100 - \Delta/2, 0, \Delta/2)$. Obviously, in the new allocation scheme 1 and 3 have increased payoffs, therefore 1 is also willing to accept the new scheme. But since the payoff for 2 becomes 0, he is dissatisfied with the new scheme and proposes $(100 - \Delta/4, \Delta/4, 0)$. 2 and 3 continue in turn like this. In the very end, the scheme is $(100, 0, 0)$. Now 1 is very satisfied, whereas 2 and 3 can no longer suggest any scheme to increase their own payoff that is acceptable to 1.

Note that even though the allocation scheme in the core is extreme, it does not mean that it definitely does not exist in reality. In many situations, there is some basis for the allocation scheme in the core.

Example 6.6 A real-life version of the Glove Game—Chinese enterprise producing high-speed rail equipment cutting each other's prices overseas ("Merger of CNR and CSR: Avoiding cut-throat competition overseas", by Chen Shanshan, *China Business News*, 28 October 2014).

In China there are only two manufacturers of high-speed rail equipment; they are China CNR Corporation Limited (CNR) and China South Locomotive & Rolling Stock Corporation Limited (CSR). The enterprises got into a 'vicious' competition.

In 2012 the government of Argentina planned to purchase urban rail trains. CNR participated in the tender, with the first quote being 2.39 million US dollars per train. Compared with overseas competitors, the price seemed rather high, but being reassured by quality, they felt very confident.

What surprised CNR was that, CSR who had never been in the Argentine market, suddenly got in for the kill During the first tender, it quoted a low price of 1.27 million US dollars per train, nearly 50% lower than CNR's. The Argentinians who put the tender out were shocked but pleased, and immediately proposed that for the second tender no enterprise could quote exceeding 1.27 million US dollars per train.

In such a situation, CNR could only quote a low 1.26 million US dollars per train in the second tender, but they never imagined that CSR's quote was even lower, at only 1.21 million US dollars per train. In the end CSR got the two tenders that it wanted.

Opposite to the situation in China, many manufacturing powers in rail and train-making usually only have one enterprise in the country for rail manufacturing, such as Germany's Siemens and France's Alstom.

6.2.6.3 Allocation Schemes that Do not Satisfy Any Conditions of the Core

For some special types of cooperative games, the 'core' might be empty. In other words, there may not be any allocation schemes that satisfy conditions of the core.

For situations where a core does not exist in cooperative games, readers can refer to 'Example 8.4' of Chap. 8 of this book.

6.3 The Shapley Value

Because the core can include many allocation schemes, the core as the solution to coalitional games is often an interval. In such a situation, the core solution (i.e. the allocation scheme to be adopted) is not unique. By comparison, the Shapley value can give a unique solution to coalitional games and therefore is widely used in management practice. The basic idea of the Shapley value is to carry out the allocation according to the players' average contribution margin for the coalition.

When players join a coalition, how much is a player's marginal contribution to the coalition is often connected with the order in which he joins the coalition, which is the formation of the order in which players join the coalition. Take the example of a two-person game $G^{\{1,2\}}$: in order that the two persons are motivated in forming a coalition, let us suppose $v(1) + v(2) < v(1,2)$, i.e. the coalition is superadditive.

First, let us take a look at the permutation $\sigma_1 = (1,2)$, i.e. 1 joins the coalition first, and 2 joins afterwards. Here, 1 joins the coalition first. At the time, because there is no other player in the coalition, then the marginal contribution 1 makes to the coalition equals the payoff he gets when he goes it alone, i.e. $v(1) - v(0)$. However, 2 joins the coalition after 1, and there is already player 1 in the coalition. Thus, the marginal contribution 2 makes to the coalition is $v(1,2) - v(1)$. Thus, in the permutation $\sigma_1 = (1,2)$, 1 and 2's marginal contribution vector $m^{\sigma_1}(v)$ for the coalition formed is:

$$m^{\sigma_1}(v) = \{[v(1) - v(0)],\ [v(1,2) - v(1)]\}$$

According to the assumption that the coalition is superadditive, i.e. $v(1) + v(2) < v(1,2)$, therefore $v(1,2) - v(1) > v(2)$. This means that because 2 joins the coalition later, his marginal contribution to the coalition is greater than the payoff he gets when he goes it alone.

Here, **the reason the marginal contribution vector is important is because during negotiations of the allocation schemes, each player's marginal contribution to the coalition is key for obtaining their own benefits from the allocation scheme**.

Next, in the permutation $\sigma_2 = (2,1)$, 1 and 2's marginal contribution vector $m^{\sigma_2}(v)$ for the coalition is:

$$m^{\sigma_2}(v) = \{[v(2,1) - v(2)],\ [v(2) - v(0)]\}$$

Likewise, $v(1) + v(2) < v(1,2)$, therefore $v(1,2) - v(2) > v(1)$. This means that now player 1, in joining the coalition later, his marginal contribution to the coalition is greater than the payoff he gets when he goes it alone.

As a result, in coalitional games $G^{\{1,2\}}$, if allocation is based on the player's marginal contribution to the coalition, then the player who joins later is more advantageous in the allocation scheme. Now, here is a problem. If players are allowed to choose the moment to join a coalition, they will choose to be the last to join.

An idea for solving this problem, is to allow each player to join the coalition in all the permutations, then add together the marginal contribution vectors of all the possible permutations to get an average value. This average value is used as the basis for each player's allocation. This is the basic idea of the Shapley value.

According to this idea, regarding $G^{\{1,2\}}$, the two players' allocation scheme formed by their respective average contribution margin vector is:

$$(x_1, x_2) = \left(\frac{[v(1) - v(0)] + [v(2, \ 1) - v(2)]}{2!}, \ \frac{[v(2) - v(0)] + [v(1, \ 2) - v(1)]}{2!} \right)$$

(6.6)

Likewise, for a coalitional game $G^{\{1,2,3\}}$ with three players, we have:

$$(x_1, \ x_2, \ x_3) = \left(\begin{array}{c} \frac{\{[v(1)(\text{The permutation is } 1, \ 2, \ 3) - v(0)] + [v(1)(\text{The permutation is } 1, \ 3, \ 2) - v(0)]\}}{+ \ \{[v(2, \ 1) - v(2)] + [v(3, \ 1) - v(3)]\} + \{[v(2, \ 3, \ 1) - v(2, \ 3)] + [v(3, \ 2, \ 1) - v(3, \ 2)]\}}{3!}, \\[2ex] \frac{\{[v(2)(\text{The permutation is } 2, \ 1, \ 3) - v(0)] + [v(2)(\text{The permutation is } 2, \ 3, \ 1) - v(0)]\}}{+ \ \{[v(1, \ 2) - v(1)] + [v(3, \ 2) - v(3)]\} + \{[v(1, \ 3, \ 2) - v(1, \ 3)] + [v(3, \ 1, \ 2) - v(3, \ 1)]\}}{3!}, \\[2ex] \frac{\{[v(3)(\text{The permutation is } 3, \ 1, \ 2) - v(0)] + [v(3)(\text{The permutation is } 3, \ 2, \ 1) - v(0)]\}}{+ \ \{[v(1, \ 3) - v(1)] + [v(2, \ 3) - v(2)]\} + \{[v(1, \ 2, \ 3) - v(1, \ 2)] + [v(2, \ 1, \ 3) - v(2, \ 1)]\}}{3!} \end{array} \right)$$

(6.7)

Similarly, the allocation in a n-person coalition can be made by the average contribution margin vector.

Definition 6.6 Game G^N's Shapley value $\phi(v)$ is:

$$\phi(v) = \frac{1}{n!} \sum_{\sigma_j \in \pi(N)} m^{\sigma_j}(v)$$

(6.8)

In the formula, $\pi(N)$ is the set of the various permutations that can be produced by each player in the set of players N. σ_j is one of the permutations in $\pi(N)$. $m^{\sigma_j}(v)$ is a marginal contribution vector formed under the permutation σ_j with n players' marginal contribution.

It can be seen from the Formula (6.8) that game G^N's Shapley value $\phi(v)$ is the vector formed by the arithmetic mean of the various players' marginal contribution vector. The Shapley value $\phi(v)$ is different from the concept of the core in that the Shapley value is an allocation scheme for n players; it is not a set of the various allocation schemes.

The Shapley value $\phi(v)$ is a vector; its i component $\phi_i(v)$ is the average value of the marginal contribution by player i in the various possible permutations in N, i.e.

$$\phi_i(v) = \frac{1}{n!} \sum_{\sigma_j \in \pi(N)} m_i^{\sigma_j}(v)$$

In the formula, $m_i^{\sigma_j}(v)$ is player i's marginal contribution in the n-player permutation σ_j.

Example 6.7 For coalitions with payoffs that are only additive (not superadditive nor subadditive), i.e. for coalition $S, \ T \in 2^N$, if $S \cap T = \emptyset$, then $v(S \cup T) = v(S) + v(T)$. Here, there is $v(S \cup \{i\}) = v(S) + v(i)$, i.e. $v(S \cup \{i\}) - v(S) = v(i)$. Clearly, it does not matter what permutation the player is in; the i marginal contribution always equals the payoff $v(i)$ he gets when going it alone. Therefore, the i component of the Shapley value is (note that for a coalition of n players, the total number of possible permutations is $n!$):

$$\phi_i(v) = \frac{1}{n!} \sum_{\sigma_j \in \pi(N)} m_i^{\sigma_j}(v) = \frac{n! \, v(i)}{n!} = v(i).$$

Thus, for this kind of coalition that is only additive, its Shapley value is $(v(1), v(2), \ldots, v(n))$.

Chapter 7
Allocating Benefits in Coalitions

7.1 Allocating System Efficiency

In management practice, most of the time the goal of cooperation is system efficiency. When the number of elements in a system does not reach a certain critical point, the system efficiency is 0, but once the number of elements reaches the critical point, system efficiency suddenly emerges.

Let us assume that the set of players is $N = \{1, 2, \ldots, n\}$, the special feature of the characteristic function for a coalition with system functionality is $v(1) = v(1, 2) = \cdots = v(1, 2, \ldots, n - 1) = 0, v(1, 2, \ldots, n) > 0$.

For instance, player 1 has land, player 2 has labour (foreman), and player 3 has technology know-how and facility. On their own, land, labour, or technological facility cannot produce efficiency. But if these three elements are brought together, an excellent payoff can be produced.

For coalitions with system efficiency, let us assume $N = \{1, 2, \ldots, n\}$, then for any player to have marginal contribution, he must wait till the n player joins the coalition. Also, when he is positioned at n, the marginal contribution is the same whatever the permutation, i.e. it equals $v(1, 2, \ldots, n)$; the marginal contribution is 0 at any other position. Therefore, the Shapley value component of player i is:

$$\phi_i(v) = \frac{1}{n!} \sum_{\sigma_j \in \pi(N)} m_i^{\sigma_j}(v) = \frac{(n-1)!}{n!} v(1, 2, \ldots, n) = \frac{v(1, 2, \ldots, n)}{n} \tag{7.1}$$

That is, for coalitions with system payoff, the allocation scheme determined by the Shapley value is for the various players to equally share in the coalition's payoff. In the above formula, there is $(n - 1)!$ in the numerator because player i has to wait till player n to join the coalition for his marginal contribution to equal $v(1, 2, \ldots, n)$. At this time, there are altogether $(n - 1)$ players who have joined the coalition before him; the number of permutations for these people is $(n - 1)!$. But in other situations, player i's marginal contribution is 0.

© China Economic Publishing House and Springer Nature Singapore Pte Ltd. 2018
S. Sun and N. Sun, *Management Game Theory*,
https://doi.org/10.1007/978-981-13-1062-1_7

7.2   The Problem of Unequal Status

7.2.1   A Landlord and Tenant Cooperative Game

7.2.1.1   The Shapley Value When Tenants Do not Form a Sub-coalition (This Example Is Taken from *Cooperative Game Theory* by Dong Baomin, Wang Yuntong and Guo Guixia)

Moulin (1988) brought forth a landlord and tenant cooperative game problem.

Let us suppose there are $n + 1$ players. Player 0 is the landlord who owns the land (but the landlord does not do labour work); players 1, 2 ..., n are the n tenants who can offer labour. Each tenant's capacity for labour is the same. Let us also assume that M is the subset of m tenants in n formed by tenants and landlord 0 (may or may not include landlord 0).

The characteristic of this cooperative coalition is:

The landlord alone cannot produce any payoff, i.e. $v(0) = 0$

Tenants without a landlord also cannot produce any payoff, i.e. $v(M) = 0$, when $0 \notin M$.

When the landlord and tenants join up to form a coalition, the payoff is an increasing function of the number of tenants, i.e. $v(M) = f(m)$, when $0 \in M$. At this point $|M| = m + 1$.

For this cooperative game, the set of players is $\{0, 1, 2, \cdots, n\}$, in which 0 is the landlord. When the landlord is positioned at $i + 1$ in the permutation, there are i tenants before him. At this time, his marginal contribution is $f(i)$. Therefore, the landlord's Shapley value is:

$$\phi_0(v) = \frac{1}{n+1} \sum_{i=1}^{n} f(i) \tag{7.2}$$

For each of the tenant, because their labour capacity is the same, the Shapley value is the same, i.e. for any tenant j, his Shapley value is the whole of the grand coalition's payoff minus the landlord's Shapley value and then equally divided by n tenants:

$$\phi_j(v) = \frac{1}{n} \left[f(n) - \frac{1}{n+1} \sum_{i=1}^{n} f(i) \right] \tag{7.3}$$

For this example, if the game's core is used to solve the allocation scheme, then it can be seen that in the core there are many allocation schemes. It includes a scheme that gives all the payoff to the landlord ($x_0 = v(0, N), x_i = 0, n \geq i \geq 1$), and a scheme that allows the tenants to take all the payoff $\left(x_0 = 0, x_i = \frac{v(0, N)}{n}, n \geq i \geq 1 \right)$, as well as the various 'compromise' schemes between the two extremes (because a core is a convex set).

7.2.1.2 The Situation When Farmers Form a Sub-coalition

With regards to this problem, Sun Shaorong points out that, to prevent the landlord using his uncommon status to get hold of more benefits, these n tenants can form a sub-coalition called tenant N. Thus, $n+1$ players are simplified to become two players: landlord 0 and tenant N. In this way, the problem evolves to become a problem of system efficiency. Because if there is only landlord 0 or tenant N (indicates that there are n tenants joining or not joining the coalition as an entity), foodstuff cannot be produced. That is $v(0) = 0$, $v(N) = 0$.

According to the Shapley value solution above for coalitions with system efficiency, we have:

$$\phi_0(v) = \frac{v(0, N)}{2} = \frac{f(n)}{2}, \tag{7.4}$$

i.e. the landlord and n tenants each side getting half of the overall payoff.

Then, with regards to the allocation among the n tenants, as their capacity for labour is the same, the payoff is divided equally for everybody, i.e.

$$\phi_i(v) = \frac{f(n)}{2n}, i = 1, 2, \ldots, n \tag{7.5}$$

Comparing Formula (7.5) with Formula (7.3), it can be proved that when $f(m)$ is a linear increasing function, the numerical result is the same for both. In other words, the two algorithms produce the same Shapley value, i.e. the tenants' payoff is the same.

However, if the factor of negotiation capability is taken into consideration, then the situation is not the same for Formulae (7.5) and (7.3). It is not difficult to see that in Moulin's landlord-tenant cooperative game problem, the landlord is obviously more dominant in the game. Because if the landlord does not provide the land, the cooperation payoff between the tenants is 0, while the landlord can choose to cooperate with different tenants. In such a situation competition between tenants emerges, leading to the tenants competing to reduce their own share. This was actually the reason why in old China some landlords could raise the rent of the land really high, so there was exploitation.

Therefore, if we follow Sun Shaorong's improvement proposal for all the tenants to form a sub-coalition, it is of advantage for the protection of the tenants' interests. Then the landlord can only negotiate with the tenants as an entity, and without the participation of the tenants, his payoff will become 0. In such a situation, the landlord's negotiation capacity equals the tenants' as a whole. From this it can be seen that farmers associations or trade unions are important in some countries.

Likewise in the human resources market, since viewed in terms of quantities the number of employers is always fewer than the number of job seekers (each enterprise counts as a player, and each job seeker also counts as a player), in the game of the employer and the job seekers, the employer is always in a dominant position. In such a situation, if the country does not take the lead in stipulating a minimum wage, employers can keep wages really low. This is the reason governments of many countries step out to stipulate a standard for minimum wage.

7.2.2   The Problem of Asymmetric Pairing

7.2.2.1   The Shapley Value When the Weak Parties Do not Form a Sub-coalition

The problem of pairing is very common in management practice. In some game theory works, pairing left and right hand gloves or left and right shoes are generally used as examples. In fact, the problem of pairing technology and capital or land and facilities are more pertinent.

The so-called problem of asymmetric pairing is in fact an imbalance of numbers on pairing the two sides, causing a dominant side and a weak side.

Let us still use the problem of pairing gloves as an example. Suppose $N = \{1, 2, 3\}$, in which 1 owns a left-hand glove, 2 and 3 each owns a right-hand glove. The value of each pair of matched glove is 100 yuan. The value of any single glove is 0.

When the extreme nature of the core as an allocation scheme was discussed before, the allocation scheme in the unique core was given as $(100, 0, 0)$ for that problem.

Now let us take a look at the Shapley value for this problem. For a situation with only three players, (this example is taken from *An Introduction to Cooperative Games* by Shi Xiquan), there are altogether only six possible permutations, and therefore all of them can be listed. The various players' marginal contribution vectors formed by their marginal contributions under the various permutations can also be listed, as shown in Table 7.1.

In the six marginal contribution vectors, player 1 got 100 four times, and player 2 and player 3 each got 100 once. Thus, the Shapley vector is $(400/6, 100/6, 100/6)$.

For the problem of pairing n left gloves with $(n + 1)$ right gloves, as n gets bigger, the difference in quantity of the two becomes smaller, the advantage of having a left

Table 7.1 Marginal contribution vector

Permutations	Marginal contribution vector (the number inside the brackets is the number of the player)
123	0, 100 (2), 0
132	0, 100 (3), 0
213	0, 100 (1), 0
231	0, 0, 100 (1)
312	0, 100 (1), 0
321	0, 0, 100 (1)

glove will become less and less. The Shapley value of the gloves game when $n = 1,000,000$: for each person holding a left-hand glove the allocation was 0.500433×100 yuan, for each person holding a right-hand glove the allocation is 0.499557×100 yuan. It can be seen that in the situation where there are 1,000,000 left-hand gloves and 101 right-hand gloves, the advantage of having a left-hand glove is basically gone.

7.2.2.2 The Shapley Value When the Weak Parties Form a Sub-coalition

Now let us reconsider the problem. $N = \{1, 2, 3\}$, 1 has a left-hand glove, and 2 and 3 each has a right-hand glove; each matched pair of gloves is valued at 100 yuan. What is different from before is that, as pointed out by Sun Shaorong, in fighting for their interests, 2 and 3 as the weak parties will form a sub-coalition, and agree in advance that, they as a sub-coalition will pair up with 1, i.e. 2 and 3 agree in advance to either join or not join the grand coalition together.

In such a situation, it evolves into a question of the Shapley value for a coalition with system efficiency; because a payoff can only be produced when 1 and the sub-coalition formed by 2 and 3 are combined as a grand coalition.

As before, according to the Shapley value solution for a coalition with system efficiency, player 1 gets half of the grand coalition's payoff, whereas the sub-coalition 23 gets the other half of the payoff, which 2 and 3 then equally share. In such a situation, the Shapley value is (50, 25, 25). As we can see, when 2 and 3 form a sub-coalition, the payoff for players 2 and 3 is improved because, if they do not form a sub-coalition, the Shapley value is $(400/6, 100/6, 100/6)$.

7.3 The Question of Sets of Cost

The question of sets of cost refers to the situation where in cooperative games when the various players work alone, their costs or payoffs can form an increasing sequence, and the cost of the person with high cost (or payoff) can offset the cost of the person with low cost.

For instance, there is a straight road that leads to town. The cost of constructing the road for the village furthest from town is higher than that for the village next furthest from town, and the cost for the village next furthest from town is higher than that for the villages nearer to town, when they each bear the construction cost alone…; for colleagues who travel the same road and share a taxi home after work, the one who lives furthest from work will have a higher cost when hiring the taxi alone than the one who lives the next furthest from work… and so on.

Below let us use the example of the Airport Game (Littlechild and Owen 1973) to illustrate the Shapley value solution for the question of cost sets (this example is taken from *Cooperative Game Theory* by Dong Baomin, Wang Yuntong and Guo Guixia).

Let us suppose that n Airline companies $N = \{1, 2, \ldots, n\}$ own different sizes of aircrafts, and each of these airlines has only a certain size of aircrafts. Large aircrafts need a longer runway than the small aircrafts. Runways that can accommodate large aircrafts to take off and land will definitely accommodate small aircrafts.

Suppose for aircraft size i, the construction cost of the runway is $c_i (i = 1, 2, \ldots, n)$. We can also suppose the size of aircraft is in this order $1 \leq 2 \leq \cdots \leq n$. Thus, the construction cost of runways for each size of aircraft alone is $c_1 \leq c_2 \leq \cdots \leq c_n$. Thus, for the coalition $S \in 2^N$ formed by some airlines, the cost outlay is only the runway construction cost for the biggest aircraft in S. Therefore, the cost borne by all airlines in coalition S is:

$$v(S) = \max\{c_p, c_q, \ldots, c_w\}, (p, q, \ldots, w) \in S \tag{7.6}$$

Let us first look at the cost borne by company 1 with the smallest model of aircraft. There are $(n - 1)!$ possible permutations for company 1 to be at the first position. When company 1 is at the first position, its marginal contribution (increased marginal cost to the coalition) is c_1. When company 1's position is second or even later in the permutation, because company 1's cost is the lowest, the runway cost of the company (companies) positioned before it already includes company 1's runway cost, so its increased marginal cost to the coalition is 0. For n companies, the number of all the permutations is $n!$, and therefore company 1's Shapley value is:

$$\phi_1 = \frac{(n - 1)!}{n!} c_1 = \frac{c_1}{n} \tag{7.7}$$

Next, let us look at the cost borne by company 2. If it is the first to join the coalition, then its marginal contribution to the coalition is c_2. There are $(n - 1)!$ permutations for this situation. For n companies, the number of all the permutations is $n!$, therefore the probability for this is $\frac{(n-1)!}{n!} = \frac{1}{n}$. Here the cost borne by company 2 should be $\frac{c_2}{n}$.

When company 2 is the second to join the coalition, if the one before it happens to be company 1, the marginal contribution to the coalition by company 2 is the portion of its cost that exceeds company 1's cost $(c_2 - c_1)$. There are $(n - 2)!$ permutations for this situation, and therefore the probability is $\frac{(n-2)!}{n!}$ for this kind of situation. The cost borne by company 2 should be $\frac{(n-2)!}{n!}(c_2 - c_1)$.

If company 2 is positioned third or even later, then there is at least one company before it whose runway cost is greater than c_2. Here, company 2's marginal contribution to the coalition is 0.

Therefore, company 2's Shapley value is

$$\phi_2 = \frac{c_2}{n} + \frac{(n - 2)!}{n!}(c_2 - c_1) = \frac{c_2}{n} + \frac{1}{(n - 1)n}(c_2 - c_1)$$

$$= \frac{c_2}{n} + \frac{n - (n-1)}{(n-1)n}(c_2 - c_1)$$

$$= \frac{c_2}{n} + \left(\frac{1}{n-1} - \frac{1}{n}\right)(c_2 - c_1) = \frac{c_1}{n} + \frac{1}{n-1}(c_2 - c_1)$$

Next, let us look at the cost borne by company 3. If it is the first to join the coalition, then its marginal contribution to the coalition is c_3. There are $(n-1)!$ permutations for this situation. For n companies, the number of all the permutations is $n!$, and therefore the probability for this is $\frac{(n-1)!}{n!} = \frac{1}{n}$. Here the cost borne by company 3 should be $\frac{c_3}{n}$.

When company 3 is the second to join the coalition, there are three kinds of situations.

The first is that the company before it happens to be company 1, and the marginal contribution to the coalition by company 3 is the portion of its cost that exceeds company 1's cost $(c_3 - c_1)$. There are $(n-2)!$ permutations for this situation, so the probability is $\frac{(n-2)!}{n!}$ for this kind of situation. Here the cost borne by company 3 is $\frac{(n-2)!}{n!}(c_3 - c_1)$.

The second is that the company before it happens to be company 2, and the marginal contribution to the coalition by company 3 is the portion of its cost that exceeds company 2's cost $(c_3 - c_2)$. There are $(n-2)!$ permutations for this situation, so the probability is $\frac{(n-2)!}{n!}$ for this kind of situation. Here the cost borne by company 3 is $\frac{(n-2)!}{n!}(c_3 - c_2)$.

The third is that at least one of the companies before it is neither company 1 nor company 2. Here, before company 3, there must be company 4 or a company of a higher number. In such a situation, company 3's marginal contribution to the coalition is 0.

All in all, when company 3 is the second to join the coalition, its cost is $\frac{(n-2)!}{n!}[(c_3 - c_1) + (c_3 - c_2)]$.

When company 3 is the third to join the coalition, there are two kinds of situations.

One is that the company before it happens to be company 1 and company 2. Here, company 3's marginal contribution to the coalition is the portion of its cost that exceeds company 2's cost $(c_3 - c_2)$. The number of permutations for this situation is $(n-3)!2!$, of which $(n-3)!$ is the number of permutations of $n-3$ companies after removing company 1, company 2 and company 3 from the n companies. $2!$ is the possible permutations of company 1 and company 2 when these two companies are in front of company 3. Therefore, the probability is $\frac{(n-3)!2!}{n!}$ for this kind of situation. Here the cost borne by company 3 is $\frac{(n-3)!2!}{n!}(c_3 - c_2)$.

The other situation is that at least one of the companies before it is neither company 1 nor company 2. Here, the companies before it is either company 4 or a company of a higher number. Since the length of runway of any company whose number is greater than 3 will be longer than company 3's runway, here, company 3's marginal contribution to the coalition is 0.

Bringing together the various sequences of company 3 joining the coalition, company 3's Shapley value is:

$$\phi_3 = \frac{c_3}{n} + \frac{(n-2)!}{n!}[(c_3 - c_1) + (c_3 - c_2)] + \frac{(n-3)!\,2!}{n!}(c_3 - c_2)$$

$$\begin{aligned}
\phi_3 &= \frac{c_3}{n} + \frac{(n-2)!}{n!}[(c_3 - c_1) + (c_3 - c_2)] + \frac{(n-3)!\,2!}{n!}(c_3 - c_2) \\
&= \frac{c_3}{n} + \frac{1}{n(n-1)}(c_3 - c_1) + \left[\frac{1}{n(n-1)} + \frac{2}{n(n-1)(n-2)}\right](c_3 - c_2) \\
&= \frac{c_3}{n} + \frac{n-(n-1)}{n(n-1)}(c_3 - c_1) + \left[\frac{1}{n(n-1)} + \frac{2}{n(n-1)(n-2)}\right](c_3 - c_2) \\
&= \frac{c_3}{n} + \frac{1}{(n-1)}(c_3 - c_1) - \frac{1}{n}(c_3 - c_1) + \left[\frac{1}{n(n-1)} + \frac{2}{n(n-1)(n-2)}\right](c_3 - c_2) \\
&= \frac{c_1}{n} + \frac{1}{(n-1)}(c_3 - c_1) + \left[\frac{1}{n(n-1)} + \frac{2}{n(n-1)(n-2)}\right](c_3 - c_2) \\
&= \frac{c_1}{n} + \frac{1}{(n-1)}(c_3 - c_1) + \left[\frac{n-2}{n(n-1)(n-2)} + \frac{2}{n(n-1)(n-2)}\right](c_3 - c_2) \\
&= \frac{c_1}{n} + \frac{1}{(n-1)}(c_3 - c_1) + \left[\frac{(n-1)-(n-2)}{(n-1)(n-2)}\right](c_3 - c_2) \\
&= \frac{c_1}{n} + \frac{1}{(n-1)}(c_3 - c_1) + \frac{1}{(n-2)}(c_3 - c_2) - \frac{1}{(n-1)}(c_3 - c_2) \\
&= \frac{c_1}{n} + \frac{1}{(n-1)}(c_3 - c_1) + \frac{1}{(n-2)}(c_3 - c_2) - \frac{1}{(n-1)}(c_3 - c_2) \\
&= \frac{c_1}{n} + \frac{1}{(n-1)}(c_2 - c_1) + \frac{1}{(n-2)}(c_3 - c_2)
\end{aligned}$$

In fact, for the Airport Game, the Shapley value of company $k(k = 1, 2, \ldots, n)$ is:

$$\phi_k = \frac{c_1}{n} + \frac{1}{n-1}(c_2 - c_1) + \cdots + \frac{1}{n-k+1}(c_k - c_{k-1}) = \sum_{i=1}^{k} \frac{c_i - c_{i-1}}{n-i+1} \quad (7.8)$$

Looking at the formula, we find that cost c_1 of the shortest runway is shared equally among the n companies. Then the cost difference between the next-shortest runway cost c_2 and the shortest runway cost c_1, which is $(c_2 - c_1)$, is equally divided by the other $(n-1)$ companies except company 1 ..., until $(c_{k+1} - c_k)$ is shared by $(n-k+1)$ companies between company k and company n. Thus, companies with longer runways bear higher cost, which is, however, lower than if that company 'goes it alone'.

Chapter 8
Coalitions—Disintegration and Stability

Coalitions in games refer to groups formed by some of the players of the game for pre-agreed actions. The aim of this kind of coalitions is to protect the interests of coalition members.

Generally speaking, forming coalitions can achieve more payoffs for the coalition as a whole, so that participating individuals can be allocated more payoffs than working alone. However, in many situations, coalitions can disintegrate.

Usually, for grand coalitions, the existence of individual rationality or small coalition rationality, or put another way, selfishness in individuals or in the small coalitions, can cause the coalition to break up, or cause cooperation within the coalition to become confrontational, leading to diminished payoffs, and in some situations can even lead to unstable game behaviour.

8.1 Individual Rationality Causes a Cooperative Game to Become Confrontational and Diminished Payoffs

The so-called individual rationality mainly refers to game players maximising his own payoff as his guiding principle when deciding his choice of actions. This kind of individual rationality often causes cooperation to become confrontational within a coalition.

Example 8.1 Disintegration of a coalition caused by individual rationality—Prisoner's Dilemma (this example is based on *Game Theory and Information Economics* by Zhang Weiying, and *Introduction to Game Theory* by Wang Zeke and Li Jie).

Prisoner's Dilemma was formulated by Merrill Flood, Melvin Dresher and Albert Tucker of Rand Corporation in 1950. A classic Prisoner's Dilemma is:

Two suspects A and B are arrested by police, but the police do not have enough evidence to convict the two. So they are imprisoned separately, and the police offered them each the following choices:

Table 8.1 Prisoner's Dilemma

	B denies	B confesses
A denies	Serve a 1-year sentence, serve a 1-year sentence	Serve a 10-year sentence, immediately released
A confesses	Immediately released, serve a 10-year sentence	Serve an 8-year sentence, serve an 8-year sentence

If someone confesses and pleads guilty whereas the other one denies, then former is immediately released, and the one who denies is sentenced to 10 years imprisonment.

If both deny, then both will be sentenced to 1 year imprisonment due to insufficient evidence.

If both confess, their crime is confirmed, but because of their confession they will both be sentenced to 8 years imprisonment (Table 8.1).

Therefore each prisoner faces two choices: Deny or confess.

The character of this game is that no matter what the other chooses, the optimal choice for each is always to confess: If the other denies and I choose to confess, I will be released; choosing to deny I will be given a one-year sentence; if the other confesses, I will be sentenced to 8 years; choosing to deny I will be given a 10-year sentence.

If both suspects are individually rational, they will both choose to confess, and the result is that each will be sentenced to 8 years.

However, this is the result if the two do not cooperate.

If the two are trusted friends, they will agree beforehand that in the event of being caught, they will both choose to deny so that they will get the lightest sentence. This kind of situation is when a cooperative coalition emerges.

8.2   Small Coalition Rationality Causes a Cooperative Game to Become Confrontational and Diminished Payoffs

Small coalition rationality has already been defined in Chap. 6 of this book. It mainly refers to the situation where some members of a grand coalition consider forming a small coalition for higher payoff. This manifestation of 'small coalition selfishness' is small coalition rationality.

In reality, there are models of sub-coalition in grand coalitions; sub-coalition rationality often harms the interests of the grand coalition.

Example 8.2 Small coalition rationality causing confrontations and diminished payoffs

Please take a look at the example of a three-person game below (this example is quoted from *Introduction to Game Theory* by Wang Zeke and Li Jie).

In the three-person game represented in Table 8.2, each player has two strategies: Player A's strategy is a_1 or a_2; player B's strategy is b_1 or b_2; player C then decides whether it is scheme One or scheme Two.

There are two pure strategy Nash Equilibria in this game (a_1, b_1, scheme One) and (a_2, b_2, scheme Two). Comparing the two equilibria, it can be seen that (a_1, b_1, scheme One) is the Pareto optimal, i.e. $v(A, B, C)$ in the situation of (a_1, b_1, scheme One) is greater than $v(A, B, C)$ in the situation of (a_2, b_2, scheme Two). Under (a_1, b_1, scheme One) the payoff vector is (0, 0, 10), i.e. player C's payoff is 10, whereas player A's and player B's payoff is 0 respectively. In this situation, player C must share some of his payoff with player A and player B. At least, A and B's payoff should be greater than 1, such as in the (2, 2, 8) allocation scheme. Otherwise, player A and player B will conspire together and respectively choose (a_2, b_2). In such a situation player C will be forced to choose scheme Two, resulting in a payoff vector of (−1, −1, 5). Obviously, when this result is compared with the suggested scheme (2, 2, 8) at the equilibrium point (a_1, b_1, scheme One) the payoff has declined for all players.

Table 8.2 The harm of Small Coalition Rationality

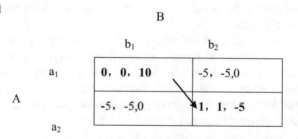

Player C chooses scheme One

Player C chooses scheme Two

If the three sides can form a coalition, then everyone will agree in advance to reach the equilibrium point (a_1, b_1, scheme One). At the same time, to console A and B, the allocation scheme (2, 2, 8) is adopted. Only in such a situation can a true equilibrium be achieved. Of course, this kind of equilibrium is not Nash Equilibrium but a game in agreement under cooperation.

8.3   Individual Rationality Causes the Disintegration of a Coalition

The following is an example of individual rationality causing a coalition to disintegrate.

Example 8.3 Garbage disposal game (this example is taken from *Introduction to Game Theory* by Wang Zeke and Li Jie).

The Indian economist Avinash K. Dixit et al. formulated the Garbage Game, constructing a coalition that can be broken up. Its main content is:

The players are the people of n households. Each household has one bag of garbage to deal with. Every time a bag of garbage is not disposed, the loss suffered by the garbage owner is 1.

In this example, m is used to indicate a m-person coalition; the characteristic function for the m-person coalition (i.e. the coalition's payoff) is $v(m)$, and $1 \leq m \leq n$ is stipulated.

Let us assume that the people of n households are not well-behaved and a fight over the garbage broke out between them: everyone wants to throw their garbage to someone else to dispose for them. In such a situation, if m people unite together, the characteristic function of this m-person coalition is $v(m) = -(n - m)$ because this coalition can throw m bags of garbage to people outside the coalition, whereas those outside the coalition can throw $n - m$ bags of garbage to people in the coalition.

From the characteristic function it can be seen that the greater the number of people in m, i.e. the nearer it is to n, the greater the $v(m)$. However, if m equals m, the coalition's characteristic function is not $v(m) = -(n - m) = -(n - n) = 0$ but $v(n) = -n$, because it assumes that the garbage cannot be thrown to people outside the n people. Here, the existence of coalition m is meaningless - it is the same as not having a coalition.

Thus, the coalition with the biggest payoff is a coalition formed by $n - 1$ persons; here $v(n - 1) = -1$: These $n - 1$ people can throw their garbage to the poor chap excluded from the coalition. The best revenge for this poor chap is to throw his bag of garbage to a person in the coalition (randomly to someone in the coalition).

However, individual rationality can easily break up this coalition.

The person excluded from the coalition can inform $n - 2$ people in the coalition, asking them to throw their garbage to the person in the coalition who has not been informed, whose bag of garbage he will take on himself. Thus, in the new situation, the $n - 2$ people who have been informed by him no longer face the risk of being in receipt of garbage. That is $v(n - 2) = 0$.

Whereas for the person originally excluded from the coalition, his characteristic function or the cost he has to bear is reduced from $v(i) = -(n - 1)$ to $v(i) = -1$. Thus, the original coalition m has disintegrated.

8.4 Small Coalition Rationality Causes the Disintegration of a Coalition

Example 8.4 Majority rule in internal allocation of benefits causes coalition disintegration (this example is edited from *An Introduction to Cooperative Games* by Shi Xiquan).

Below let us discuss an example in a cooperative game where small coalition rationality leads to the disintegration of the coalition.

Three players 1, 2, 3 form a coalition. The wealth they all own together is valued at 1. This coalition has a rule for allocating wealth: The allocation scheme is decided by a vote, and the majority rule is adopted in voting.

According to this allocation rule, anyone who works alone (i.e. has a difference of opinion to others for allocating wealth) will be allocated 0.

That is $v(1) = 0, v(2) = 0, v(3) = 0$.

Anyone who holds the same opinion as the majority (two or three people) can take possession of all of the wealth valued at 1.

That is $v(1, 2) = 1, v(2, 3) = 1, v(1, 3) = 1, v(1, 2, 3) = 1$.

Obviously for this coalition the imputation set is:

$$I(v) = \{x_1, x_2, x_3|, x_1 + x_2 + x_3 = 1, x_1 \geq 0, x_2 \geq 0, x_3 \geq 0,\}$$

Now let us prove that in the imputation set there are no subsets that satisfy small coalition rationality, i.e. the core in this imputation set is an empty set.

There are three separate situations to prove this.

The element (i.e. allocation scheme) in this imputation set must be one of these three situations:

Situation 1: The three players are all allocated a certain share of the assets, i.e. $x_1 > 0, x_2 > 0, x_3 > 0$.
Situation 2: Two of the three players are allocated a share of 0 of assets.
Situation 3: One of the three players is allocated a share of 0 of assets.

Now we can prove that these three situations do not satisfy small coalition rationality, i.e. they are not the allocation schemes in the core.

Let us first look at situation 1. In this situation, any two of the three persons can form a small coalition to share the wealth valued at 1. Since the payoff increases when two persons share the wealth valued at 1 rather than three persons sharing (here each of the three must get a certain share of the wealth), therefore there is no small coalition rationality in situation 1.

Next, let us look at situation 2. If there are two persons without any payoff, i.e. their payoff is 0, then they can definitely form a sub-coalition in order to overthrow the voting in situation 2. Therefore, situation 2 is not in the core either.

Lastly let us look at situation 3. This is when one person's payoff is 0; for the other two, the payoff for one is $\Delta \neq 0$, and for the other it is $1 - \Delta$. For convenience, let us assume that the imputation is $(0, \Delta, 1 - \Delta)$. Now, player 1 can propose an improved allocation scheme $(\Delta - \frac{\Delta}{2}, 0, 1 - \Delta + \frac{\Delta}{2})$. Here, player 3 will also agree this new allocation scheme because his payoff has also increased. Thus a sub-coalition of player 1 and player 3 is formed. Similarly, player 2 is unwilling to accept the loss and will propose a new allocation scheme $(0, \Delta - \frac{\Delta}{2} - \frac{\Delta}{4}, 1 - \Delta + \frac{\Delta}{2} + \frac{\Delta}{4})$, which starts up the sub-coalition of player 2 and player 3. And so it continues. It can be seen that situation 3 is not a stable result either, and it is not in the core.

It can be proved that for a coalition formed by any players of odd number greater than 3, so long as it adopts the majority rule, it is unstable.

This example shows that, for decision-making in collectives to deal with issues such as allocating internal benefits and assets, the institution of vote in the collective currently used widely, which features decision based on the opinion of the majority, is unsuitable. Here is a real life example. In the 1990s, the faculty of a college in metallurgy voted to decide the last 'staff accommodation allocation scheme'. Since the majority of teaching staff is male in a college in metallurgy, the voting outcome was 'only male teachers will be allocated accommodation'. The incident caused dissatisfaction among female teaching staff, which was reported to the provincial Women's Federation and was only resolved when the Women's Federation intervened.

8.5   A Coalition's Stability

8.5.1   Confrontations and a Coalition's Stability

A coalition's stability is decided by individuals' satisfaction degrees to distribution schemes in a coalition.

From the perspective of a coalition's stability, in addition to meeting the requirements of the core, the coalition's payoff allocation scheme should also prevent any individual in the coalition from proposing improvement schemes sharing with others against a certain player. An improvement scheme of this kind is called a confrontation against a certain player's original allocation scheme.

If a new improvement scheme is proposed directed at the confrontation, i.e. a confrontation aimed at the confrontation, this is called a counter-confrontation. Obviously, if the person proposing the confrontation can foresee that his confrontation will be denied by a counter-confrontation, then the confrontation will not be proposed. An allocation scheme is stable if it does not face confrontations, and it is the very basis of a stable coalition.

Example 8.5 An example of a confrontation and counter-confrontation (this example is taken from *An Introduction to Cooperative Games* by Shi Xiquan).

Let us consider a three-person game, $v(1, 2, 3) = 100$, $v(1) = 0$, $v(2) = 0$, $v(3) = 0$, $v(2, 3) = 50$, $v(1, 2) = 100$, $v(1, 3) = 100$. In this game, player 1's status is relatively important. If players 2 or 3 can cooperate with player 1, the coalition's payoff can reach 100. But if 2 and 3 cooperate, then the payoff can only be 50.

From this, someone proposes an imputation of $(75, 25, 0)$, i.e. player 1 and 2 are united. As 1 is more important, he gets 75; player 2 has at least 25, because if 2 gets less than 25, he can form an alliance with 3 to equally share the payoff of 50 and can get at least 25.

However, it is very likely that player 1 will disagree with the scheme $(75, 25, 0)$ because he can also form an alliance with 3, and achieves imputation $(76, 0, 24)$.

Thus, $(76, 0, 24)$ is player 1's confrontation against player 2.

But with regards to $(76, 0, 24)$ player 2 can also propose a counter-confrontation $(0, 25, 25)$. Player 3 will clearly support this counter-confrontation because player 3's payoff will be bigger.

In such a situation, player 1 expects that his proposed confrontation $(76, 0, 24)$ towards $(75, 25, 0)$ will be counteracted by $(0, 25, 25)$ proposed by player 2, leading to an even worse result for him. Therefore, player 1 will not confront $(75, 25, 0)$.

However, will player 2 propose the confrontation $(0, 27, 23)$ against $(75, 25, 0)$? We find that with regards to this confrontation, player 1 can propose a counter-confrontation $(75, 0, 25)$. This scheme will also be supported by player 3. Obviously, this counter-confrontation is even more disadvantageous for player 2.

Therefore, player 2 will not propose a confrontation against $(75, 25, 0)$.

Therefore, if only players 1 and 2 are considered, then the imputation $(75, 25, 0)$ is stable—neither party will propose a confrontation scheme.

Similarly, if only players 1 and 3 are considered, the imputation $(75, 0, 25)$ is also a stable imputation because the status of player 2 and player 3 are the same in the cooperative game.

However, for imputation $(75, 25, 0)$, if we consider player 3 exists, then it is not a stable scheme because player 3 can propose $(76, 0, 24)$ directed against player 1, and can also propose $(0, 26, 24)$ directed against player 2, so as to break up the coalition of 1 and 2.

Likewise, if only players 2 and 3 are considered, then $(0, 25, 25)$ is also a stable imputation. It is because, if player 2 proposes the confrontation $(0, 26, 24)$, then player 3 will propose a counter-confrontation $(75, 0, 25)$.

However, if we consider player 1, then $(0, 25, 25)$ is not a stable imputation either. It is because player 1 can propose $(74, 26, 0)$ against player 2, or he can propose $(74, 0, 26)$ against player 3, so as to break up the coalition of 2 and 3.

Bringing together the above, in this cooperative game, the sub-coalitions $S_1 = \{1, 2\}$, $S_2 = \{1, 3\}$, and $S_3 = \{2, 3\}$ are not stable coalitions because it is not possible to find an imputation that cannot be broken up by confrontations.

8.5.2 Definitions of Confrontation and Counter-Confrontation

For the two imputations x and y, and also with regards to any coalition:

$$x >_{(S)} y \text{ If and only if } x_i > y_i, \forall i \in S \tag{8.1}$$

Definition 8.1 Regarding coalition S and the imputation x, the definition

$$e(S, x) = v(S) - \sum_{i \in S} x_i \tag{8.2}$$

Is the excess for S regarding x.

The excess $e(S, x)$ is the difference between sub-coalition S's payoff and the sum total of all the coalition members' payoff from the imputation. For all the members in S, the sum of the payoff is $\sum_{i \in S} x_i$ from the imputation x. Coalition S can produce a payoff of $v(S)$. Thus, if the excess is $e(S, x) \geq 0$, it shows that coalition S can produce a payoff of $v(S)$ and it is not 'completely depleted'. In other words, the smaller the excess, the more the S members are willing to accept x.

Confrontation can be defined as follows:

Definition 8.2 A confrontation by player i against another player j and the imputation x is expressed as (y, S), of which y is another imputation, $S \subseteq N, i \in S, j \notin S$, (y, S) satisfies $e(s, y) = 0$ as well as $y >_{(S)} x$.

In Example 8.3, the confrontation initiated by player 1 against player 2 and the scheme is $((76, 0, 24), S = \{1, 3\})$. $(76, 0, 24)$ is better for player 1 and 3 than $(75, 25, 0)$. At the same time the excess of $\{1, 3\}$ regarding $(76, 0, 24)$ is 0.

Definition 8.3 Player j's counter-confrontation against player i and his confrontation (y, S) is expressed as (z, T). z is another imputation, and T is another sub-coalition, of which $j \in T, i \notin T$, and also $T \cap S \neq \phi$, (z, T) satisfies: $e(T, z) = 0, z \geq_{(T)} y$.

For Example 8.3, regarding player 1's confrontation $((76, 0, 24), S = \{1, 3\})$, player 2 proposes his counter-confrontation $z = (0, 25, 25), T = \{2, 3\}$, so that the payoffs for players 2 and 3 are increased.

8.6 Stability of Game Behaviour in Non-cooperative Games

Let us now analyse the question of stability of game behaviour in non-cooperative games. In some situations, if the non-cooperative game can reach Nash Equilibrium, then the game behaviour of the various players in that game reaches a state of stability. However, if the behaviour of everybody changes continuously, then the game behaviour is unstable.

The economist Harold Hotelling pointed out in 1929 that if two manufacturers A and B produce the same good, then the closer the geographical locations for the sale of goods from these two manufacturers, the stronger their substitutability. Consumers who are further away from the geographical locations where the goods are sold, their cost for the purchase is higher. Therefore customers will choose the nearest shops to buy products. Therefore, for manufacturers producing the same product, they are only in competition with manufacturers in the same neighbourhood. Let us assume that there are only two manufacturers in the market. Each manufacturer only has one point of sale and can only be distributed along a 1 km line segment. Parallel to this at a certain distance is where consumers are evenly distributed along a 1 km straight line. As in Fig. 8.1 (this example is from *Introduction to Game Theory* by Wang Zeke and Li Jie).

Here, what is a better choice for the point of sale for each manufacturer?

It is often assumed that this line segment is equally divided into quarters, with the first manufacturer's point of sale A at 1/4, the second manufacturer's point of sale B at 3/4. Indeed, if the two manufacturers are in a rather good relationship, and also both sides abide by their agreement, then it is possible to distribute the point of sale for both sides in this way.

But from the perspective of competitive games, if A moves along to the right, then A will have more customers, and B will have fewer customers. In such a situation, B will also move along to the left to recapture lost customers.

Thus, A continues to shift to the right, and B continues to shift to the left. In the end, both sides reach close to the middle of the segment at position 1/2 (Fig. 8.2). Now, neither side can move along anymore, and so equilibrium is reached.

However, if along that line segment there are three manufacturers' points of sale A, B, C. What is the situation? The conclusion is that in this situation, there is no stable equilibrium: these points of sale will continue to move.

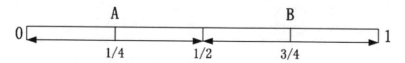

Fig. 8.1 The distributing of the point of sale if they are in good relationship

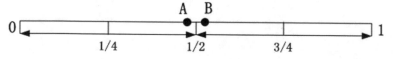

Fig. 8.2 The distributing of the point of sale if they are in competition

Chapter 9
Bottom Line for Negotiations and Solutions

9.1 Overview of Negotiations and Negotiation Proportional Models

9.1.1 An Overview of Negotiations

Negotiation means that in situations of opposing interests, the two sides carry out talks to fight for their own interests. As to opposing interests for the two sides in games, if one party's interests increase, then the other party's interests will reduce.

From the point of view of a negotiation outcome, if both sides reach an agreement, then the negotiation is a success. By contrast, if both sides cannot agree in the end, then the negotiation has broken down.

Generally speaking for both parties, the payoff for a successful negotiation is better than when a negotiation breaks down. In other words, an expectation from both sides for success in negotiations is an important basis for conducting the negotiation. On the other hand, without breaking off the negotiation, both sides will work to increase their own benefits. Therefore, negotiations are a process of struggle in a cooperative situation.

9.1.2 Negotiation Proportional Models

To define negotiation from the perspective of mathematical description, let us assume that when A and B are successful in a negotiation, the total benefits for both sides is 1. Also, let us assume that the negotiation process for both sides is to fight for a proportion of benefits as great as possible during the benefits allocation. Let us suppose that through negotiations, the proportion of benefits A gets is $0 < x < 1$, then the proportion of benefits B gets is $0 < 1 - x < 1$; therefore the negotiation

© China Economic Publishing House and Springer Nature Singapore Pte Ltd. 2018
S. Sun and N. Sun, *Management Game Theory*,
https://doi.org/10.1007/978-981-13-1062-1_9

outcome is $(x, 1 - x)$. If the negotiation breaks down, then the outcome is $(0, 0)$. Therefore, as long as $x \neq 0, x \neq 1$, a successful negotiation is better than a failed negotiation. Therefore, any outcome in $(x, 1 - x) (0 < x < 1)$ is Nash Equilibrium. The problem is, to determine the numerical specifics of $0 < x < 1$ that fulfils this condition, both sides have to fight given their terms and capacity.

The negotiation outcome is connected to the bottom line they set for themselves prior to the negotiation. The so-called bottom line is the least acceptable settlement set by the player himself prior to the event. From a mathematical perspective, the negotiation bottom line is a set formed by the total payoff from the various allocation schemes after a successful negotiation. We use BL_A to indicate A's bottom line and BL_B to indicate B's bottom line. For example, both sides determine in advance that the bottom line in the negotiation is that their own payoff cannot be less than one half of the overall payoff, i.e. for A it is $BL_A = x \geq \frac{1}{2}$, whereas for B it is $BL_B = 1 - x \geq \frac{1}{2}$. In such a situation, only $x = \frac{1}{2}$ will be an acceptable outcome. In such a situation, the proportion of payoff for both sides is $\frac{1}{2}$.

The fundamental condition for a successful negotiation is that the intersection of sets formed by the bottom line of both sides is not empty. That is $BL_A \cap BL_B \neq \varnothing$.

For instance, $BL_A = \{x \geq \frac{1}{3}\}$, $BL_B = \{1 - x \geq \frac{1}{4}\}$, then $BL_A \cap BL_B = \{x = [\frac{1}{3}, \frac{3}{4}]\}$. In such a situation, both sides can haggle over $x = [\frac{1}{3}, \frac{3}{4}]$ to fight for an advantageous outcome for themselves.

Figure 9.1 is an image representation of $BL_A \cap BL_B = \{x = [\frac{1}{3}, \frac{3}{4}]\}$. In this graph, the horizontal axis is A's proportion of payoff x, the vertical axis is B's payoff $1 - x$, to the right of the dotted line $x = \frac{1}{3}$ is $BL_A = \{x \geq \frac{1}{3}\}$, and above the dotted line $1 - x = \frac{1}{4}$ is $BL_B = \{1 - x \geq \frac{1}{4}\}$. The solid line in Fig. 9.1 is the outcomes set of the negotiation; it is located at the top right corner of the point of intersection of the dotted lines $x = \frac{1}{3}$ and $1 - x = \frac{1}{4}$, and indicates the viable outcomes set of the negotiation.

If $BL_A = \{x \geq \frac{2}{3}\}$, $BL_B = \{1 - x \geq \frac{1}{2}\}$, then $BL_A \cap BL_B = \varnothing$. In such a situation, the negotiation between the two sides will definitely break down.

Figure 9.2 shows how $BL_A = \{x \geq \frac{2}{3}\}$ and $BL_B = \{1 - x \geq \frac{1}{2}\}$ lead to the situation of $BL_A \cap BL_B = \varnothing$. To the right of the dotted line $x = \frac{2}{3}$ is $BL_A = \{x \geq \frac{2}{3}\}$, and above the dotted line $1 - x = \frac{1}{2}$ is $BL_B = \{1 - x \geq \frac{1}{2}\}$. It can be seen from the graph that the negotiation outcomes set is the solid line in Fig. 9.2, located at the lower left corner of the point of intersection of the dotted lines $x = \frac{2}{3}$ and $1 - x = \frac{1}{2}$, and indicates that there is no viable outcomes set for this negotiation.

In some negotiations, the set of viable outcomes is only a point, such as $BL_A = x \geq \frac{1}{2}$ and $BL_B = 1 - x \geq \frac{1}{2}$. In such a situation, only $x = \frac{1}{2}$ is a mutually acceptable outcome, i.e. the elements in $BL_A \cap BL_B$ is only a point. The graph is represented in Fig. 9.3.

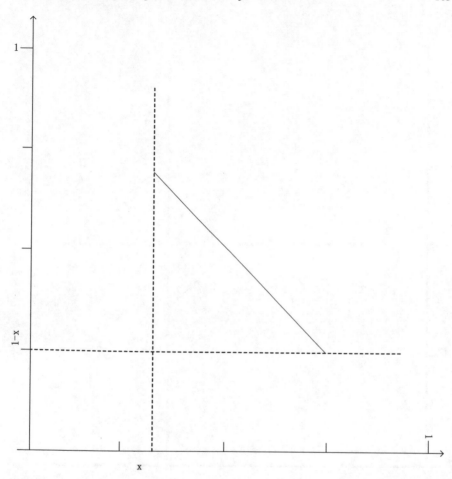

Fig. 9.1 The situation at $BL_A \cap BL_B = \{x = [\frac{1}{3}, \frac{3}{4}]\}$

9.2 A General Model for Negotiations and an Objective Bottom Line

9.2.1 A General Model for Negotiations

Let us now discuss a general model for negotiation by two sides.

Let us suppose the two sides are A and B in the negotiation.

If the two sides are successful in the negotiation, it will produce a total payoff of $v(A, B)$.

If the negotiation breaks down, A's payoff is $v(A)$, and B's payoff is $v(B)$.

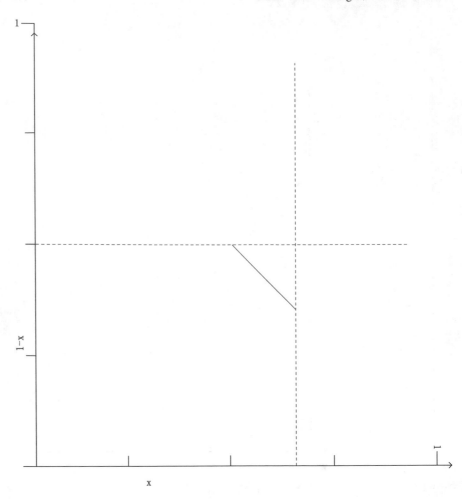

Fig. 9.2 The situation at $BL_A \cap BL_B = \varnothing$

Obviously, $v(\mathsf{A}, \mathsf{B}) > v(\mathsf{A}) + v(\mathsf{B})$ is a fundamental condition for holding the negotiation.

Let us suppose that the negotiation is successful; A's payoff is A_p, and B's payoff is $B_p = v(\mathsf{A}, \mathsf{B}) - A_p$.

9.2.2   The Objective Bottom Line in a Negotiation

The so-called objective bottom line means that if a negotiation breaks down, one party can still achieve a minimum payoff. If, for that one party, its payoff in a successful negotiation is not even as high as a failed negotiation, then a successful negotiation

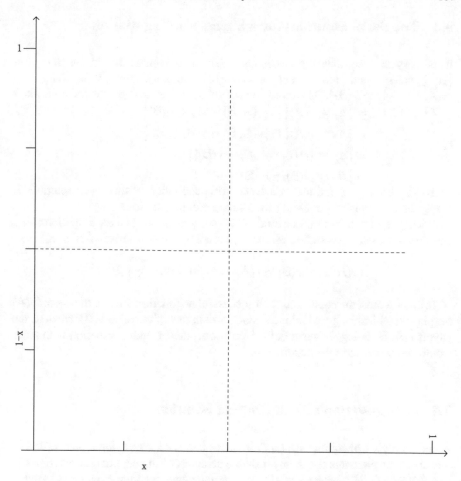

Fig. 9.3 The situation at $BL_A \cap BL_B = \frac{1}{2}$

becomes meaningless. Therefore, in the event that a negotiation breaks down, the minimum payoff that one side can obtain is called its objective bottom line in the negotiation.

It can be seen from the analysis above that A's objective bottom line for the negotiation is $BL_A = \{A_p \geq v(\mathsf{A})\}$, and B's is $BL_B = \{v(\mathsf{A},\mathsf{B}) - A_p \geq v(\mathsf{B})\}$, i.e. $BL_B = \{v(\mathsf{A},\mathsf{B}) - v(\mathsf{B}) \geq A_p\}$. In other words, B's objective bottom line only allows the other side to have a payoff not exceeding $v(\mathsf{A},\mathsf{B}) - v(\mathsf{B})$.

In the actual negotiation process, the bottom lines set by both sides are often much higher than the objective bottom lines.

9.3   The Set of Solutions for a Successful Negotiation

It is easy to see that the acceptable set of outcomes for A is $BL_A = \{A_p \geq v(A)\}$, and the acceptable set of outcomes for B is $BL_B = \{v(A, B) - A_p \geq v(B)\}$. Therefore the set of outcomes acceptable for both sides

is

$$
\begin{aligned}
BL_A \cap BL_B &= \{A_p \geq v(A)\} \cap \{v(A, B) - A_p \geq v(B)\} \\
&= \{A_p \geq v(A)\} \cap \{v(A, B) - v(B) \geq A_p\} \\
&= \{A_p = [v(A), v(A, B) - v(B)]\} \\
&= \{B_p = [v(B), v(A, B) - v(A)]\}
\end{aligned}
$$

This is the set of solutions for a successful negotiation. If this set of solutions is empty, then there is no possibility for a successful negotiation.

If a negotiation is treated as a kind of cooperative game, then the conditions for a possible successful negotiation can be expressed by a characteristic function, i.e.:

$$
v(A, B) - v(B) \geq v(A), \ v(A, B) - v(A) \geq v(B)
$$

In other words, the conditions for a successful negotiation are that the payoff $v(A)$ for player A when he goes it alone is smaller than that of player B $v(B)$, whereas the payoff $v(A, B)$ is bigger when both sides cooperate. In such a situation both sides would be interested in the negotiation.

9.4   A Negotiation's Nash Product Solution

A non-empty set of solutions $BL_A \cap BL_B$ is often not a set of single points. Within this are many elements (i.e. many possible schemes of the negotiation outcomes in Fig. 9.1 $BL_A \cap BL_B$ form a straight line, among which are infinite element 'points', each point being in fact a negotiation outcome; of course sometimes elements are limited in the solution set for a problem under negotiation).

In such a situation, the final negotiation outcome is determined by the negotiation capability of the two sides. Therefore, theoretically speaking, only a 'set' of various outcomes can be predicted for the negotiation, but it is impossible to accurately give a unique negotiation outcome.

However, for a solution based on an 'overall optimal set', then the negotiation outcome is often unique. For instance, the Nash product solution.

Let us assume that the general model for a problem under negotiation is: The two parties in the negotiation are A and B. If the negotiation is successful, the total payoff produced is $v(A, B)$; if the negotiation fails, A's payoff is $v(A)$, and B's payoff is $v(B)$. Let us suppose that the negotiation is successful, A's payoff is A_p, and B's payoff is $B_p = v(A, B) - A_p$.

Assuming that A's utility is u_A, B's utility is u_B, then the Nash product is defined as the product of the increment of the payoff $u_A(A_p - v(A))$ achieved by A's success in the negotiation and the utility increment $u_B(B_p - v(B))$ achieved by B's success in the negotiation:

$$u_A(A_p - v(A))u_B(B_p - v(B)) \tag{9.1}$$

When each of A's and B's utility increment equals their payoff increment (this situation is common, i.e. when neither A nor B has special economic requirement for the size of the payoff, and also the payoff range under consideration is small enough to not cause an obvious diminishing or increasing marginal utility), the Nash product is:

$$[A_p - v(A)][B_p - v(B)] = [A_p - v(A)][v(A, B) - A_p - v(B)]$$
$$= [B_p - v(A)][v(A, B) - B_p - v(A)] \tag{9.2}$$

In the situation that a non-empty set of solutions $BL_A \cap BL_B$ has many elements, we can seek the biggest value of its Nash product. Clearly, this is a unique solution and it satisfies the 'optimal collective principle'.

Example 9.1 Asymmetrical utility function as a solution for the negotiation

The two sides in the negotiation have different utility functions. In this, party A has to repay a debt of 550,000 yuan or else his house worth of 1,000,000 yuan will be repossessed. If it can be guaranteed that the house will not be repossessed, then the utility increases by five units. Therefore, his utility function jumps up a point at 500,000, i.e. it is a linear function with jumps, as shown in Fig. 9.4.

A's utility function is:

$$U_A = \begin{cases} 0.05A_p, & A_p < 550,000 \text{ yuan} \\ 5 + 0.05A_p, & A_p \geq 550,000 \text{ yuan} \end{cases}$$

As B does not have the above special situation, therefore his utility function is just a general linear function. However, he places more importance to economic payoff, as shown in Fig. 9.5.

B's utility function is:

$$U_B = 0.1B_p$$

If the negotiation is successful, then the total payoff produced is $v(A,B) = 100,0000$ yuan; if the negotiation fails, A's payoff is $v(A) = 50,000$ yuan when going it alone, B's payoff is $v(B) = 70,000$ yuan when going it alone.

Let us suppose that the negotiation is successful, A's payoff is A_p, then B's payoff is $B_p = v(A, B) - A_p$.

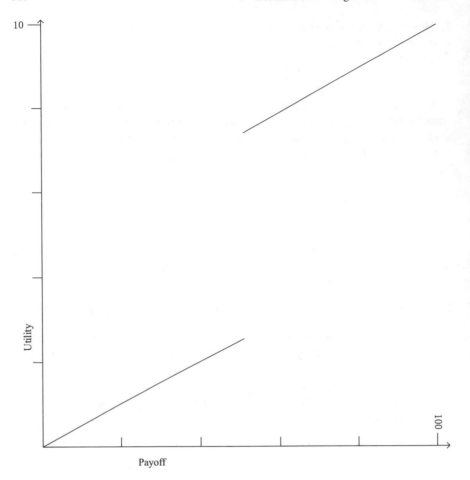

Fig. 9.4 Linear utility functions with jumps

Considering A and B's utility function described above, the Nash product for this negotiation problem is:

$$u_A(A_p - v(A))u_B(B_p - v(B)) = 0.05(A_p - v(A))0.1(B_p - v(B))$$
$$= [0.05(A_p - v(A))][0.1(v(A, B) - A_p - v(B))]$$
$$= [0.05(A_p - 5)][0.1(100 - A_p - 7)], \text{When } A_p < 550,000$$

$$u_A(5 + A_p - v(A))u_B(B_p - v(B)) = 0.05(5 + A_p - v(A))0.1(B_p - v(B))$$
$$= [0.05(5 + A_p - v(A))][0.1(v(A, B) - A_p - v(B))]$$
$$= [0.05(5 + A_p - 5)][0.1(100 - A_p - 7)], \text{When} A_p \geq 550,000$$

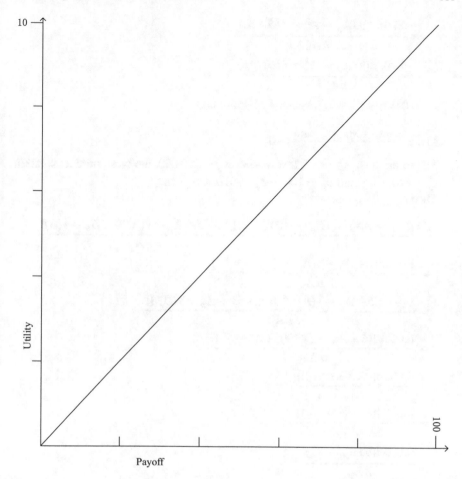

Fig. 9.5 Continuous linear utility function

When $A_p < 550{,}000$, Nash product derived function is:

$$\frac{du_A(A_p - v(\mathbf{A}))u_B(B_p - v(\mathbf{B}))}{dA_p} = \frac{d\{[0.05(A_p - 5)][0.1(100 - A_p7)]\}}{dA_p}$$

$$= \frac{0.005d[(A_p - 5)(100 - A_p - 7)]}{dA_p}$$

$$= \frac{0.005d[(A_p - 5)(93 - A_p)]}{dA_p}$$

$$= \frac{0.005d[(A_p(93 - A_p) - 5(93 - A_p)]}{dA_p}$$

$$= \frac{0.005d(93A_p - A_p^2 - 465 + 5A_p)}{dA_p}$$

$$= \frac{0.005d(98A_p - A_p^2 - 465)}{dA_p}$$

$$= 0.005(98 - 2A_p), \text{ When } A_p \geq 550,000$$

Make $\frac{du_A(A_p - v(A))u_B(B_p - v(B))}{dA_p} = 0$

The solution is: $A_p^* = 49$, that is when $A_p < 550,000$, the Nash product solution is $A_p^* = 490,000$ yuan, $B_p^* = 100 - A_p^* = 510,000$ yuan.
Then $A_p \geq 550,000$:

$$\frac{du_A(A_p - v(A))u_B(B_p - v(B))}{dA_p} = \frac{d\{0.05[5 + A_p - v(A)]0.1[B_p - v(B)]\}}{dA_p}$$

$$= \frac{d\{0.05[5 + A_p - v(A)]0.1[B_p - v(B)]\}}{dA_p}$$

$$= \frac{d\{0.05(5 + A_p - v(A))]0.1(v(A, B - A_p - v(B)))]\}}{dA_p}$$

$$= \frac{0.005d[(5 + A_p - 5)(100 - A_p - 7)]}{dA_p}$$

$$= \frac{0.005d[A_p(93 - A_p)]}{dA_p}$$

$$= \frac{0.005d[A_p(93 - A_p)]}{dA_p}$$

$$= \frac{0.005d(93A_p - A_p^2)}{dA_p}$$

$$= 93 - 2A_p, \text{ When } A_p \geq 550,000$$

Make $\frac{du_A(A_p - v(A))u_B(B_p - v(B))}{dA_p} = 0$

The solution is: $A_p^* = 46.5$, that is, when $A_p^* \geq 550,000$, the Nash product solution is $A_p^* = 465,000$ yuan, $B_p^* = 100 - A_p^* = 535,000$ yuan.
Since the independent variable domain of that function is $A_p^* \geq 550,000$, and also take into account that $\frac{du_A(A_p - v(A))u_B(B_p - v(B))}{dA_p} = 93 - 2A_p$ is a linear function, whereas the Nash product solution is $A_p^* = 465,000$ yuan $< 550,000$ yuan, it can be determined that when it is $A_p = 550,000$, the Nash product achieves the highest value. To substitute $A_p = 550,000$ for $A_p \geq 550,000$ in the Nash product:

$$u_A(5 + A_p - v(A))u_B(B_p - v(B)) = 0.05(5 + A_p - v(A))0.1(B_p - v(B))$$

$$= [0.05(5 + A_p - v(A))][0.1(v(A, B) - A_p - v(B))]$$

$$= [0.05(5 + A_p - 5)][0.1(100 - A_p - 7)]$$

$$= 0.005 A_p(93 - A_p)$$
$$= 0.005 \times 55(93 - 55)$$
$$= 10.45$$

Then when $A_p < 550{,}000$, $A_p^* = 49$, substituted into the Nash product at this stage, the highest value achieved is:

$$u_A(A_p - v(\mathbf{A}))u_B(B_p - v(\mathbf{B})) = 0.05(A_p - v(\mathbf{A}))0.1(B_p - v(\mathbf{B}))$$
$$= [0.05(A_p - v(\mathbf{A}))][0.1(v(\mathbf{A}, \mathbf{B}) - A_p - v(\mathbf{B}))]$$
$$= [0.05(A_p - 5)][0.1(100 - A_p - 7)]$$
$$= [0.005(A_p - 5)(93 - A_p)$$
$$= [0.005(49 - 5)(93 - 49)$$
$$= 9.68$$

Clearly, the Nash product 9.68 ($A_p = 490{,}000$) is smaller than the Nash product $10.45(A_p = 550{,}000)$. Therefore, the Nash product solution scheme for that negotiation is: For the cooperative payoff of 1,000,000 yuan, side A gets 550,000 yuan, and side B gets 450,000 yuan.

Chapter 10
Evolution and Stability

10.1 Evolutionarily Stable Point

In usual circumstances, the object of behaviour management is groups with large numbers of non-rational individuals. Owing to the existence of widespread irrationality and sheep-flock effect, the analysis of individual rationality with game theory is often not in accord with reality. In fact, these non-rational individuals usually cannot correctly make a one-off, optimal choice in their game behaviour. Instead, through simple imitating and learning, they continuously adjust their own behaviour until it eventually reaches optimal utility. This is the evolutionary game.

Example 10.1 Pairing game (this example is edited from *Economic Game Theory* by Xie Shiyu).

Table 10.1 shows a pairing game. Let us suppose that there is a group with a large number of individuals. Everyone makes a living by doing a certain type of work and that work can only be done by two persons pairing up (such as to lift heavy items for transporting). Thus, in this game, every player can choose two actions: to 'pair up' and to 'not pair up'. The rule for payoff is that only when both sides choose to 'pair up' can the work be effective, and then each can achieve a payoff of one unit. As long as the other chooses to 'not pair up' the pairing cannot happen and thus the work is ineffectual, and the payoff for both sides is 0. Note that we have assumed that the game is a static game, i.e. both sides choose the action simultaneously.

Table 10.1 The pairing game

Player 1	Player 2		
		Pair up	Not pair up
	Pair up	1,1	0,0
	Not pair up	0,0	0,0

© China Economic Publishing House and Springer Nature Singapore Pte Ltd. 2018
S. Sun and N. Sun, *Management Game Theory*,
https://doi.org/10.1007/978-981-13-1062-1_10

It is easy to see that in that game there are two pure Nash Equilibrium actions (pair up, pair up) and (not pair up, not pair up). Of these, (pair up, pair up) is more Pareto optimal. Therefore, if the two players are completely rational, clearly both sides will choose to 'pair up'.

Now let us suppose that the players do not have the full capacity for judgement, or that they lack the information needed for making the accurate judgement. Therefore, they can only improve their own actions continuously by trials and reviewing lessons learned from experience.

Thus, for each individual, they can meet others who may choose to 'pair up', or may also choose to 'not pair up'. Therefore, a player's payoff is connected with what type of player his is and what type of player he will meet through stochastic pairing.

Let us assume that in the group, the proportion of players who are inclined to 'pair up' is x, then the proportion of players who are inclined to 'not pair up' is $1 - x$. The payoff for players inclined to 'pair up' is u_x, and the payoff for players inclined to 'not pair up' is u_{1-x}, then:

$$u_x = x \cdot 1 + (1 - x) \cdot 0 = x$$
$$u_{1-x} = x \cdot 0 + (1 - x) \cdot 0 = 0$$

Thus, in the group, the average payoff for all the members is

$$\bar{u} = x \cdot u_x + (1 - x)u_{1-x} = x^2$$

It can be seen that the payoff for the players who are inclined to 'pair up' is higher than that of the players who are inclined to 'not pair up'. Thus, through continuous repetitions, all players will discover this difference in payoff. In other words, players who do not pair up will gradually find that it is more advantageous to change their original behaviour, and so they start to imitate the other type of players.

Thus, in the actual process the proportion of the two types of players x and $1 - x$ varies with time.

With regards to the rate of change for the players inclined to 'pair up', it is determined by two factors: One is the base figure of the 'pair up' type players—the larger the base figure, the faster the rate of change. The other is the difference between the payoff for the 'pair up' type of players and the average payoff—the bigger the difference the faster the rate of change.

Let us assume that the rate of change of the 'pair up' type players is proportional to these two factors, then the differential equation is:

$$\frac{dx}{dt} = x(u_x - \bar{u}) \tag{10.1}$$

Here, x is the proportion of the 'pair up' type of players, and $(u_x - \bar{u})$ is the difference between the payoff for the 'pair up' type of players and the average payoff. $\frac{dx}{dt}$ is the rate of change with time for the 'pair up' type of players. That equation is called an equation for the speed of evolution for groups with a large number of individuals.

When the 'pair up' players' expected payoff and the average payoff for all the players in the group are brought into the equation described above for the speed in learning, we have:

$$\frac{dx}{dt} = x(x - x^2) = x^2(1 - x) = x^2 - x^3$$

Observing the above equation, we find that if $x = 0$, i.e. initially in the group, there is no one who wants to pair up, and the group will never have people who adopt the 'pair up' behaviour. This is because learning and imitation require objects to imitate and learn from. $x = 0$ means that there simply is no object from whom imitation can be learnt. Therefore none of the players will change from their original behaviour.

When $x > 0$, i.e. in the initial state of the group when 'pair up' behaviour is already adopted by players, if the payoff for these players exceeds the average payoff, then in group the rate of change of x (the proportion of players who 'pair up') is positive, i.e. there is a gradual increase in players to 'pair up'.

The question is what the final state is with this kind of change in a group. To solve this problem, it is necessary to look at the phase diagram of the equation for the speed of evolution (Fig. 10.1).

According to the phase diagram described above of the equation for the speed of evolution, with the exception that the number of 'pair up' people is 0 in the group, the ultimate outcome evolved is that all players will be 'paired up', i.e. $x = 1$. Therefore, $x^* = 0$ and $x^* = 1$ are the two end points of the direction of evolution (the final state).

The question of stability in the final state should be considered in an evolutionary game, i.e. if an individual deviates from the state of stability for any reason, can the

Fig. 10.1 Phase diagram of the equation for the speed of evolution in a pairing game

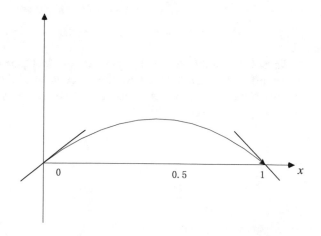

group still return to its original stable state? This is a question of how robust is the final evolved state of the group.

Let us still use the example of a pairing game. Let us assume that we are in a state when the behaviour of all players converges towards 'pairing up', there appears a proportion ε of players who, for various reasons, deviate from the equilibrium point and choose to 'not pair up'. Thus, the proportion of players who choose to 'pair up' is $1 - \varepsilon$.

Therefore, the expected payoff for the players who 'pair up' and 'do not pair up' respectively and the average payoff for the players are:

$$u_x = (1 - \varepsilon) \cdot 1 + \varepsilon \cdot 0 = 1 - \varepsilon$$
$$u_{1-x} = (1 - \varepsilon) \cdot 0 + \varepsilon \cdot 0 = 0$$
$$\bar{u} = (1 - \varepsilon) \cdot u_x + \varepsilon \cdot u_{1-x} = (1 - \varepsilon)^2$$

$u_x = 1 - \varepsilon > 0$, therefore the expected payoff for the players who 'do not pair up' is lower than that of the 'pair up' players. Thus, players who 'do not pair up' will gradually evolve to become paired up, and the final outcome is still for all players to 'pair up'. That is, $x^* = 1$ is an evolutionarily stable point (ESS). It can still remain stable even disturbed.

However, the other final state $x^* = 0$ described in the above equation for the speed of evolution, i.e. when all the players adopt the behaviour to 'not pair up', is not an evolutionarily stable point.

This is because although when in the group $x^* = 0$, no one will deviate from that point, once the group deviates from that point, it is impossible to return to it.

Let us assume that ε proportion of players deviate from 'not pair up' to 'pair up', and the expected payoff for the 'pair up' players and the average payoff for the group are:

$$u_x = \varepsilon \cdot 1 + (1 - \varepsilon) \cdot 0 = \varepsilon$$
$$u_{1-x} = \varepsilon \cdot 0 + (1 - \varepsilon) \cdot 0 = 0$$
$$\bar{u} = \varepsilon \cdot u_x + (1 - \varepsilon) \cdot u_{1-x} = \varepsilon^2$$

Thus, because the expected payoff for the 'not pair up' players is lower than the payoff for the 'pair up' players, during the process of continuous replication, the number of 'not pair up' players will decrease continuously, until the group converges to $x^* = 1$, i.e. all players 'pair up'.

10.2 Mathematical Conditions for an Evolutionarily Stable Point

An evolutionarily stable point (ESS) is a robust point against perturbation. Therefore, for ESS x^*, apart from having 0 as its equilibrium point in the evolution speed, in the event that the proportion of types in the group deviates from that point, it can also enable the proportion of types x in the group to 'automatically' return to x^*.

Let us consider the conditions for an evolutionarily stable point (ESS) from the principles of mathematics. Firstly, with the evolutionarily stable point the speed of evolution $\frac{dx}{dt} = 0$; at the same time, when x is lower than x^* because of disturbance, the speed of change $\frac{dx}{dt}$ of the proportion of types x should be greater than 0 (thus x will gradually increase with time); when x is greater than x^* because of disturbance, $\frac{dx}{dt}$ should be less than 0 (thus, x will gradually decrease with time). Thus, the condition, described with the characteristic of a second derivative, is that the derivative of $\frac{dx}{dt}$ with respect to t is less than zero, or $\frac{d\left(\frac{dx}{dt}\right)}{dt} < 0$, i.e. the slope of a tangent line in the graph is smaller than 0.

Therefore, the conditions for an evolutionarily stable point (ESS) are:

$$\begin{cases} \frac{dx}{dt} = 0 \\ \frac{d\left(\frac{dx}{dt}\right)}{dt} < 0 \end{cases} \tag{10.2}$$

In Fig. 10.1, $x = 0$ is not a evolutionarily stable point because on this point, although the evolution speed is $\frac{dx}{dt} = 0$, its slope of the tangent line is greater than 0. Only at $x = 1$ is the slope of the tangent line smaller than 0. Therefore, $x = 1$ is the evolutionarily stable point.

10.3 Follow-the-Crowd Game—the Evolutionary Equilibrium Point and the Stable Point of Evolutionary Equilibrium

The bandwagon effect in games reflects the phenomenon of a fickle public in evolutionary games.

The nature of the follow-the-crowd game belongs to coordination games. Coordination games are an important type of games in game theory. Jasmina Arifovic (2000), Hans Carlsson, Mattias Ganslandt (1998) and Paul G. Straub (1995) have all conducted a large amount of research in coordination games.

It is generally believed that there are two basic characteristics in coordination games.

Firstly, there are many Nash Equilibria in that game; secondly these Nash Equilibria can be ranked according to Pareto optimality.

According to Vincent P. Crawford and Hans Haller (1990), if each of the player in a coordination game has the same belief (such as aiming for maximum utility), and everybody has a correct expectation for the choice of action by other players (i.e. they know that the other side will also choose to maximise utility), then although there are multiple Nash Equilibria in a coordination game, there exists a unique solution, which is Pareto optimal.

Example 10.2 Follow-the-crowd game—pairing game with penalty

In order to explain the phenomenon of following the crowd, or the 'bandwagon effect' in an evolutionary game, Sun Shaorong has designed a 'pairing game with penalty'. Let us assume that in a business group of many enterprises, each enterprise can coordinate its production with other enterprises, and it can also carry out production independently. If it works with other enterprises, then both sides can exchange components to reinforce each other's advantage. Therefore the payoff for both sides is the highest, at 15 units.

If both sides choose 'independent production', then the payoff is 10 units.

If an enterprise chooses to work with another enterprise to produce products, whereas that enterprise chooses independent production, then the enterprise that chooses to 'coordinate' can only sell its products cheaply and barely achieve breakeven, i.e. its payoff is 0. However, the enterprise that chooses 'independent production' is at the receiving end of revenge by the enterprise that originally wants to 'coordinate', and so the payoff for both sides is somewhat less than if the two sides independently produce, at 5 units.

The game's payoff matrix is as shown in Fig. 10.2.

Let us assume that in the group, the proportion of players who 'coordinate' is x, then the proportion of 'independent' players is $1 - x$. The payoff for the player who chooses to 'coordinate' is u_x, and the payoff for the 'independent' player is u_{1-x}, then:

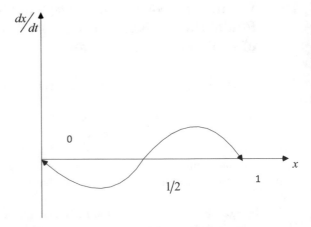

Fig. 10.2 The evolutionarily stable point and evolutionary equilibrium point in coordination games

$$u_x = x \cdot 15 + (1 - x) \cdot 0 = 15x$$
$$u_{1-x} = x \cdot 5 + (1 - x) \cdot 10 = 10 - 5x$$

Thus, in the group, the average payoff for all the members is

$$\bar{u} = x \cdot u_x + (1 - x)u_{1-x} = x \cdot 15x + (1 - x)(10 - 5x) = 15x^2 + 10 - 5x - 10x + 5x^2$$
$$= 20x^2 - 15x + 10$$

According to Formula (10.1), we have:

$$\frac{dx}{dt} = x(u_x - \bar{u}) = x(15x - 20x^2 + 15x - 10)$$
$$= x(30x - 20x^2 - 10)$$

The solution has three equilibrium points: $x^* = 0$, $x^* = 1$ and $x^* = \frac{1}{2}$.

The phase diagram for that game's equation for the speed of evolution is as shown in Fig. 10.2.

We find that in Fig. 10.2, both $x^* = 0$ and $x^* = 1$ are the evolutionarily stable points, whereas $x^* = \frac{1}{2}$ is not the evolutionarily stable point but only an unstable evolutionary equilibrium point.

From this we know that in that business group, if at a given moment the proportion of enterprises choosing to 'coordinate' x is at interval $\left(0, \frac{1}{2}\right)$, the ultimate point of the evolution will be for all enterprises to choose 'independent production', i.e. $x^* = 0$. By contrast, if at a given moment, the proportion of enterprises choosing to 'coordinate' x is at interval $\left(\frac{1}{2}, 1\right)$, the ultimate point of the evolution is that all enterprises will choose to 'coordinate', i.e. $x^* = 1$.

This outcome states that, if the parameters for a pairing game with penalty is as shown in Table 10.2, then that game is in effect a game with the bandwagon effect, i.e. when there are more enterprises in a business group that choose to 'coordinate' (when the proportion exceeds one half), for an individual enterprise, the payoff expectation is relatively high when it chooses to 'coordinate'. Therefore the stable point of evolutionary equilibrium is for all enterprises to choose to 'coordinate'. However, when there are more enterprises in a business group that choose to be 'independent' (here the proportion of enterprises that choose to 'coordinate' is less than one half), for an individual enterprise the payoff expectation is relatively high when it chooses to be 'independent'. Therefore a further stable point of evolutionary equilibrium is for all enterprises to choose to 'coordinate'. However, when by coincidence, the amount of enterprises in that business group that choose to 'coordinate' is the same as the amount of those choose to be 'independent' (each being half), then for any enterprise, the expected payoff is the same whether they choose to 'coordinate' or to be 'independent'. (When the number of enterprises is very large, it can be taken that the choice of action of a single enterprise does not alter the proportion of enterprises making the choice). The only thing is that this point is unstable. As soon as the proportion of choice is disturbed and deviates from one half, the equilibrium outcome will begin to shift towards one of the two previous situations.

	Player 2		
Player 1		Independent	Coordinate
Independent		10, 10	5, 0
Coordinate		0, 5	15, 15

Table 10.2 The bandwagon effect in a game—pairing game with penalty

This game shows that with some conflicting behaviours, where there are effective mechanisms for penalties, in groups with a large number of individuals people tend to behave the same way: either everybody is uncooperative, or everybody is cooperative.

10.4   Evolutionarily Stable Point in a Central-Tendency Game—the Hawk-Dove Game

The Hawk-Dove game is another classic game problem. Its characteristics are the opposite to the 'follow-the-crowd game'; its evolutionarily stable point is a point at the middle of the number line for the proportion of player types.

The Hawk-Dove game describes a game in which a group of homogeneous individuals fight for fixed benefits. Each individual has a choice of two behaviours; the 'hawk' refers to tough, unyielding behaviour, while the 'dove' refers to behaviours for compromise.

Since the individuals are homogeneous in that group, if players who adopt the 'hawk' behaviours meet other players who also adopt the 'hawk' behaviours, then the probability to win is 50%, and if they win they get the fixed benefit v (such as territorial gain). There is also a 50% probability for losing, and if they lose, they suffer a loss of c (such as injury in fighting).

If players adopting the 'hawk' behaviours meet players who adopt the 'dove' behaviours, then the probability for the former winning is 100%, and they will get the fixed benefits v.

If players adopting the 'dove' behaviours meet other players who also adopt the 'dove' behaviours, then both sides have an equal share of the payoff, i.e. they get $\frac{v}{2}$.

If players adopting the 'dove' behaviours meet players who adopt the 'hawk' behaviours, then they yield their payoff to the other side. Because they do not carry out fighting they do not lose anything, i.e. their total payoff is 0.

According to the description above, the Hawk-Dove game is as shown in Table 10.3.

Let us assume that in the group, the proportion of players who choose the 'hawk' behaviour is x, then the proportion of 'dove' players is $1 - x$. The payoff for players who choose the 'hawk' is u_x, and the payoff for players who choose the 'dove' is u_{1-x}, then:

$$u_x = x \cdot \frac{v-c}{2} + (1-x) \cdot v = x\frac{v}{2} - x\frac{c}{2} + v - xv = v - x\frac{v}{2} - x\frac{c}{2}$$

Fig. 10.3 The Hawk-Dove game—the evolutionarily stable point and the evolutionary equilibrium point in the central-tendency game

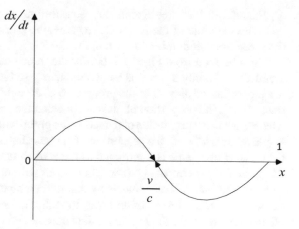

$$u_{1-x} = x \cdot 0 + (1 - x) \cdot \frac{v}{2} = \frac{v}{2} - x \cdot \frac{v}{2}$$

Thus, in the group, the average payoff for all the members is

$$\bar{u} = x \cdot u_x + (1 - x)u_{1-x} = x(v - x\frac{v}{2} - x\frac{c}{2}) + (1 - x)(\frac{v}{2} - x \cdot \frac{v}{2})$$

$$= xv - x^2\frac{v}{2} - x^2\frac{c}{2} + \frac{v}{2} - x \cdot \frac{v}{2} - x\frac{v}{2} + x^2 \cdot \frac{v}{2}$$

$$= \frac{v}{2} - x^2\frac{c}{2}$$

According to Formula (10.1), we have:

$$\frac{dx}{dt} = x(u_x - \bar{u}) = x(v - x\frac{v}{2} - x\frac{c}{2} - \frac{v}{2} + x^2\frac{c}{2})$$

$$= x(\frac{v}{2} + x^2\frac{c}{2} - x\frac{v}{2} - x\frac{c}{2})$$

$$= x(x^2\frac{c}{2} - x(\frac{v}{2} + \frac{c}{2}) + \frac{v}{2})$$

$$= \frac{x}{2}(x^2c - x(v + c) + v)$$

Table 10.3 The payoff matrix in a Hawk-Dove game (this example is edited from *Economic Game Theory* by Xie Shiyu)

	Player 2		
Player 1		Hawk	Dove
	Hawk	$\frac{v-c}{2}, \frac{v-c}{2}$	$v, 0$
	Dove	$0, v$	$\frac{v}{2}, \frac{v}{2}$

The solution has three evolutionary equilibrium points: $x^* = 0, x^* = 1$ and $x^* = \frac{v}{c}$.

In this example of the Hawk-Dove game, the phase diagram of the equation for the speed of evolution is as shown in Fig. 10.3.

It can be seen from Fig. 10.3 that for the three equilibrium points $x^* = 0$, $x^* = 1$ and $x^* = \frac{v}{c}$, only $x^* = \frac{v}{c}$ is an evolutionarily stable point. In other words, when the proportion of players in the group who choose the 'hawk' behaviour is smaller than $x^* = \frac{v}{c}$, the proportion of players choosing the 'hawk' behaviour will increase; whereas when the proportion of players in the group who choose the 'hawk' behaviour is greater than $x^* = \frac{v}{c}$, the proportion of players choosing the 'hawk' behaviour will decrease. If this game is interpreted as a mixed strategy game, then for each player the probability for choosing 'hawk' behaviour will move nearer to $x^* = \frac{v}{c}$.

Since the 'hawk' behaviour is by nature a combative behaviour, as soon as the combat is lost, the loss is often great; though winning a combat can bring payoff, when compared with the loss in a failed combat, it is often a small gain. That is to say, usually, $0 < v < c$, i.e. $0 < \frac{v}{c} < 1$. Therefore, for this game's stable point of evolutionary equilibrium, $0 < x^* = \frac{v}{c} < 1$. That is why this game is also called the central-tendency game, i.e. the proportion of players who choose the 'hawk' behaviour or the 'dove' behaviour (or the probability of each player choosing the 'hawk' or the 'dove') is often between 0 and 1. Especially if $c = 2v$, then the stable point of evolutionary equilibrium is $x^* = \frac{1}{2}$, i.e. the 'hawk' and the 'dove' each make up one half.

This game shows that for some minor conflicts (such as taking advantage of something), when effective mechanisms for penalties are lacking, there is a certain proportion in society of those individuals who are tough and unyielding and those behave peacefully, but it will not be a situation where everybody is either tough and unyielding or peaceful.

10.5   Evolutionarily Stable Point and the Evolutionary Equilibrium Point in the Layabout Game

The croaking frog game is a classic game model (this example is edited from *Economic Game Theory* by Xie Shiyu). The game is mainly concerned with whether male frogs choose to croak or not. The key feature of that game is: croaking attracts female frogs but also brings cost (consuming energy); no cost is involved in not croaking but no female frogs are attracted. A male frog that does not croak have to be a 'free rider' by relying on the croaking from other frogs, thus lowering the probability of meeting other female frogs.

To make the content of that game more relevant as a real management issue, the author has adapted this classic game so it becomes a game about whether members of a society choose to work for a living or not work but rely on social security. Of course, all the members of the groups in this game are homogeneous, i.e. each member possesses certain capability for work. Therefore, they can choose to work or to rely on social security. It is not a society where the young must work whereas the elderly people can only rely on social security.

The content of the layabout game is as follows.

Let us assume that there is a society formed by two healthy and strong young people. They can choose either to work or not work. These two people are 'layabouts', and because of their lazy nature, neither likes to work.

The problem is that, if neither works, then this 'two-person society' has no wealth whatsoever. In such a situation, the income for both is 0.

If only one person works out of the two, then it produces a small amount of labour output f. Now, the worker gets a wage of w_f ($w_f < f$). At the same time, there is a cost in working c, and therefore the worker's actual payoff is $w_f - c$. The other person who does not work can only rely on social security. As social security is equal to the reward for labour minus the wage, therefore it is $f - w_f$. As the nature of providing relief, there is $f - w_f < w_f$, i.e. $\frac{f}{2} < w_f$.

If both work, then they produce a large amount of labour output b. Now, each worker gets a wage of w_b, with an actual payoff of $w_b - c$.

Thus the two players' payoff matrix is as in Table 10.4.

If the two players are completely rational people, then an equilibrium outcome for the layabout game is determined by its parameters.

When $w_f - c < 0$, i.e. $w_f < c$, then neither player will choose to work. Thus (not work, not work) is the equilibrium point of that game.

When $w_f - c > 0$, i.e. $w_f > c$, and $w_b - c < f - w_f$, then the layabout game has two pure strategy Nash Equilibria, which are (work, not work) and (not work, work) respectively.

When $w_b - c > f - w_f$, then 'both work' is an equilibrium outcome, i.e. (work, work).

Now, let us broaden the situation to a group with numerous people. Following a stochastic pairing in the group, they play the layabout game. Let us also suppose that none of the players in that group are rational, i.e. it is not possible to form an equilibrium point from an analysis of the payoff matrix. An evolutionary equilibrium is reached only after continuous trials and errors, and reviewing past experience and lessons.

Let us assume that in the group the proportion of players who choose to 'work' is x, then the proportion of players who 'do not work' is $1 - x$. The payoff for players who choose 'to work' is u_x, the payoff for players who choose to 'not work' is u_{1-x}, then:

Table 10.4 The layabout game

		Player 2	
Player 1		Work	Not work
	Work	$w_b - c, w_b - c$	$w_f - c, f - w_f$
	Not work	$f - w_f, w_f - c$	$0, 0$

$$u_x = x(w_b - c) + (1 - x)(w_f - c) = xw_b + (1 - x)w_f - c$$
$$u_{1-x} = x(f - w_f) + (1 - x) \cdot 0 = x(f - w_f)$$

Thus, in the group, the average payoff for all the members is

$$\bar{u} = x \cdot u_x + (1 - x)u_{1-x} = x(xw_b + (1 - x)w_f - c) + (1 - x)(x(f - w_f))$$
$$= x[(xw_b + (1 - x)w_f - c) + (1 - x)f - (1 - x)w_f]$$
$$= x[xw_b - c + (1 - x)f]$$

According to Formula (10.1), we have:

$$\frac{dx}{dt} = x(u_x - \bar{u}) = x(xw_b + (1 - x)w_f - c - x[xw_b - c + (1 - x)f])$$
$$= x(xw_b + (1 - x)w_f - c - x^2w_b + xc - x(1 - x)f)$$
$$= x[x(1 - x)w_b + (1 - x)w_f - (1 - x)c - x(1 - x)f]$$
$$= x(1 - x)[xw_b + w_f - c - xf]$$
$$= x(1 - x)[x(w_b - f) + w_f - c]$$

The solution has three evolutionary equilibrium points: $x^* = 0$, $x^* = 1$ and $x^* = \frac{w_f - c}{f - w_b}$.

When $0 < \frac{w_f - c}{f - w_b} < 1$, $x^* = \frac{w_f - c}{f - w_b}$ is also an equilibrium point. Now, it is known from $0 < \frac{w_f - c}{f - w_b} < 1$, here $w_b - c < f - w_f$, i.e. when both of the two people work and the actual payoff for each person (i.e. the payoff after deducting cost) is less than the income he gets from social security when doing no work, then in the group, an equilibrium point is reached with some people working, others living on social security. Here, $x^* = 0$ and $x^* = 1$ are only evolutionary equilibrium points, not evolutionarily stable points; the only evolutionarily stable point is $x^* = \frac{w_f - c}{f - w_b}$ (Fig. 10.4).

It can be seen from Fig. 10.4 that at $0 < \frac{w_f - c}{f - w_b} < 1$, the proportion of workers in the group will remain stable at $x^* = \frac{w_f - c}{f - w_b}$. That is, if the proportion of workers exceeds $x^* = \frac{w_f - c}{f - w_b}$, then the workers suffer a loss (the expected payoff for the workers is less than the expected payoff for those living on social security), and the number of people choosing to work will gradually reduce, while the number of people choosing to live on social security will gradually increase. If the proportion of workers is less than $x^* = \frac{w_f - c}{f - w_b}$, it is not worthwhile for those living on social security; the number of people choosing to be workers will gradually increase, and the number of people choosing to live on social security will gradually reduce.

When $w_b - c > f - w_f$, i.e. when both sides work and each person's actual payoff is greater than the income of living on social security, there are two equilibrium points $x^* = 0$ and $x^* = 1$ in the layabout game, of which $x^* = 1$ is the evolutionarily stable point, i.e. an evolutionarily stable outcome is everybody working (Fig. 10.5).

In the layabout game, when $w_f - c < 0$, it is clear that the workers' payoff is less than the cost. Here, there are two equilibrium points $x^* = 0$ and $x^* = 1$ in the layabout game, of which $x^* = 0$ is the evolutionarily stable point, i.e. the evolutionarily stable outcome is nobody working (Fig. 10.6).

Fig. 10.4 The evolutionary equilibrium point and evolutionarily stable point in the layabout game (when $0 < \frac{w_f - c}{f - w_b} < 1$)

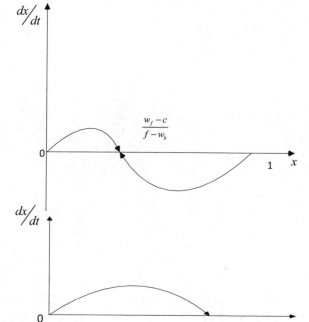

Fig. 10.5 The evolutionary equilibrium point and evolutionarily stable point in the layabout game (when $w_b - c > f - w_f$)

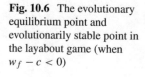

Fig. 10.6 The evolutionary equilibrium point and evolutionarily stable point in the layabout game (when $w_f - c < 0$)

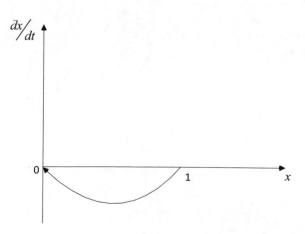

The layabout game illustrates that in given social conditions (in this example i.e. when $0 < \frac{w_f - c}{f - w_b} < 1$) the proportion of different types of members (such as layabouts and workers, law abiding citizens and citizens not abiding by the law, etc.) will remain stable at a certain point. This is in fact the basis of evolutionary games of the long-standing phenomenon that there will always be 'good people' and 'bad people' in every country, region, or organisation. Of course, as a manager, changing this condition appropriately can also change their proportions.

Bibliography

Baomin, Dong, Yuntong Wang, and Guixia Guo. 2008. *Cooperative Game Theory*. China: China Market Press.

John Maynard Smith. 2008. *Evolution and the Theory of Games*. [Translated by Pan Chunyang]. Fudan University Press. [UK].

Jorgen W. Weibull. 2006. *Evolutionary Game Theory*. [Translated by Wang Yongqin]. Shanghai People's Publishing House. [Sweden].

Littlechild SC., and Owen G. 1973. A Simple Expression for the Shapley Value in a Special Case [J]. *Management Science* 20: 370–372.

Moulin H. 1988. *Axioms of Cooperative Decision Making[M]*. Cambridge University Press.

Shi Xiquan. 2012. *An Introduction to Cooperative Games*. Peking University Press.

Wang Zeke., and Li Jie. 2010. *Introduction to Game Theory*. China Renmin University Press.

Xie Shiyu. 2002. *Economic Game Theory*. Fudan University Press.

Yue'e, Liu, Zhang Yang, Yang Jian'an, Zhang Jianhua, and Han Xiucheng. 2007. The current status in the exploitation of higher education institution patents—an investigation and reflections. *R&D Management* 19 (01): 112–118.

Zhang Weiying. 1996. *Game Theory and Information Economics*. Shanghai People's Publishing House.

© China Economic Publishing House and Springer Nature Singapore Pte Ltd. 2018
S. Sun and N. Sun, *Management Game Theory*,
https://doi.org/10.1007/978-981-13-1062-1

Printed in the United States
By Bookmasters